Astronomers' Univer

For further volumes:
http://www.springer.com/series/6960

Martin Beech

Alpha Centauri

Unveiling the Secrets of Our Nearest Stellar Neighbor

Martin Beech
Campion College
The University of Regina
Regina, Saskatchewan, Canada

ISSN 1614-659X ISSN 2197-6651 (electronic)
ISBN 978-3-319-09371-0 ISBN 978-3-319-09372-7 (eBook)
DOI 10.1007/978-3-319-09372-7
Springer Cham Heidelberg New York Dordrecht London

Library of Congress Control Number: 2014951361

Printed on acid-free paper

Springer is part of Springer Science+Business Media (www.springer.com)

The book is dedicated with appreciation to musician Ian Anderson and to all of the assembled minstrels and vagabonds that have contributed to the making of Jethro Tull. Their numerous musical adventures have accompanied, enthralled, and challenged me through many years of listening, and for this, I am grateful.

Preface

A Call to Adventure!

If there is a defining trait of being human, then a need for adventure and a desire to know what lies beyond the horizon's sweeping arc must surely be it. We yearn for adventure, be it within the confines of a favorite book, our hometown, or on some distant exotic island or mountain range.

Adventure! It fills us with passion. It provides us with a reason for action, it builds character, it shakes our assumptions, and it warms us with a sense of achievement. Scottish philosopher and Victorian essayist Thomas Carlyle once defined history as being the distillation of rumor, but surely it could better be described as the collective sum of numerous adventures, the comingled expression of journeys made by mind, body, and soul.

Adventure, it has also been said, brings out the best in us. By gritting our teeth, we have triumphed over adversity, and we assimilate wisdom. Slightly more than 100 years ago now, just within the time span of living memory, such teeth-gritting mettle saw Roald Amundsen and his Norwegian compatriots first set foot on Earth's South Pole (it contemporaneously saw the glorious death of Robert Falcon Scott and his companions). It was the same grit and determination that saw New Zealander Edmund Hillary and Nepalese Sherpa Norgay Tenzing scale the snow-clad summit of Mount Everest, the top of the world, for the very first time in 1953. It was to be only 7 years after that first great ascent before the deepest depths of Earth's oceans, the Mariana Trench, were first plumbed by Don Walsh and Jacques Piccard aboard the bathyscaphe *Trieste*.[1] Above, below, and all around – humans have literally experienced, perhaps only briefly in many circumstances, all of the topology that Earth has to offer.

Historically, high adventure has been confined to Earth and its atmosphere. This all changed, of course, not quite 50 years ago with the initiation of the American Apollo space program, which ultimately saw Neil Armstrong and Buzz Aldrin first walk upon the Moon's surface on July 21, 1969. Human beings, however, have gone no further into space than the Moon. Only robots and spacecraft (proxy human bodies made of aluminum and plastic) have continued the pioneering exploration of the planets and the deep probing of the Solar System. And yet, for all of humanity's technological skills, no spacecraft has to date reached interstellar space.[2] *Voyager 1*, the current long-distance record holder launched in 1977, is now some 18.5 billion kilometers away from the Sun, but this is a minuscule step compared to the 7.4 trillion kilometers outer radius of the Oort Cloud boundary – the zone that gravitationally separates out the Solar System, our current stomping ground, from the rest of the galaxy.

Ever hungry for adventure and raging against the yawning abyss of interstellar space, humanity has long dreamed of traveling to the stars. There may be no reasonable way of achieving such adventure in the present day or even in foreseeable decades, but the journey will assuredly begin one day; we are made of stardust, and to the stars we shall eventually make our way. But where to first? The galaxy is unimaginably large and the potential pathways innumerable. Surely, however, the first steps to the stars will be

---

[1] Remarkably, as of this writing, four times more people have walked on the surface of the Moon (12 in total – a.k.a. *The Dusty Dozen*) than have seen the ocean floor of the Mariana Trench in situ (3 in total). And although the Apollo program lasted less than 10 years, the human exploration of the deepest abyssal plain has already occupied more than half a century of adventure. The 1960 descent of the bathyscape *Trieste* was the first dive to carry a human cargo to the abyssal depths of the Challenger Deep, and then, 52 years later – on March 6, 2012 – film director and National Geographic explorer James Cameron, ensconced in the *Deepsea Challenger* submersible, descended the depths to once more cast human eyes over the floor of the Mariana Trench.

[2] I am using here the gravitational boundary, rather than the edge of the heliosphere, where the solar wind pushes up against the interstellar medium. In spite of what you may have otherwise read in press releases, the *Voyager 1* spacecraft is still very much inside of our Solar System.

via our nearest stellar neighbors, and in this case α Centauri offers up a bright and welcoming beacon.

Why α Centauri Beckons

Fortuitously close by galactic standards, α Centauri is not so remote that all hope falters at the thought of one day exploring its new-worldly domain. Not only this, but there is much about α Centauri that will be familiar to future travelers – even to our own eyes if we could be somehow transported there this very instant. Firstly, it would appear to our visual senses that we had not moved at all, for indeed, the very night sky constellations would be the same. Remarkably, as we ultimately explore α Centauri and even the solar neighborhood beyond it, the ancient zodiacal configurations will both follow us and anchor us to the deep past, and they will continue to remind us from where the journey first began. Indeed, the memory of our natal domicile will be written bright upon the sky as the Sun, as seen from α Centauri, will become a new star in the constellation of Cassiopeia.[3]

Certainly, once having arrived at α Centauri, the presence of two progenitrix stars would be odd to our sense of heritage, but these two stars up close are barely different from our familiar Sun. Indeed, they illustrate what the Sun could so easily have been, and they bookend with respect to their physical characteristics what the Sun will become in about a billion years from now.[4]

An instantaneous trip to α Centauri today would not only whisk us through a great cavern of space, it would also transport us something like 10,000 centuries into the Sun's future. Remarkably, therefore, the present-day study of α Cen A and B helps us understand the deep-time and innermost workings of our Sun. Not only this, as we shall see later on in the text, the fate and demise of life in our Solar System will be mirrored at almost the very same epoch three to four billion years hence by any life forms that

---

[3] Not only will the Sun appear as a new star in Cassiopeia, it will also be the brightest star in that constellation, far outshining Schedar (α Cassiopeia), the erstwhile brightest member as seen from Earth.

[4] It is estimated, as will be seen later, that the α Centauri system formed about six billion years ago.

might have evolved in habitable niches within the α Cen AB system. The possible worlds of α Centauri will certainly be different from those familiar to us in the Solar System, and yet they share a common future. It is an astounding testament to human ingenuity and human intellectual adventure that we can see such connections and describe them with some fair degree of confidence.

For all of its familiarity, however, there is more to the story of α Centauri than its galactic closeness at the present epoch – indeed, it is a rare closeness, and we are fortunate that it is so near at the very time that humanity can realistically envisage the launch of the very first interstellar spacecraft. Look into any modern astronomy textbook, and one of the most remarkable facts that you will discover is that our Milky Way Galaxy contains at least 200 billion stars. The Sun is far from being a lonely wonderer in space. For all its great multitude of companions, however, the Sun's existence is by and large a solitary one. Only rarely do individual stars pass close by each other, and at the present time the nearest star system, the α Centauri system, is about 28 million solar diameters away. Indeed, for stars in general there is a lot of wiggle room before any really close interactions between distinct pairs takes place. The distance between the Sun and α Centauri is still decreasing, but the two will never approach to a margin at which any distinct gravitational interaction will take place. They are indeed the astronomical equivalent of Longfellow's two passing ships in the night. But these passing ships have formed a special bond cemented by human awareness; they sail in consort, and for a brief, lingering, galactic moment they offer humanity the chance of stellar adventure and dramatic change. These passing ships afford future humans, our descendents, the incredible chance of not only finding unity in cause but of becoming cosmic voyagers – new sailors, perhaps even ambassadors, plying the interstellar sea.

Perhaps surprisingly for all of the galactic nearness of α Centauri to the Sun, there is much that we do not know or understand about its component stars; there are indeed deep and fundamental questions (thought adventures) that astronomers have yet to answer. Even at the most fundamental level, it is not presently clear if the α Centauri system is composed of two gravitationally bound stars or three. As we shall see in the main body of the text, it has long been known that the bright naked-eye α Centauri star is actually a binary system composed of two Sunlike analogs: α Cen A and α Cen B. So much is beyond doubt. What is presently unclear, however, is whether Proxima Centauri, the actual closest star to the Sun at the present epoch, forms a gravitationally bound triple system with α Centauri AB – technically, therefore, making Proxima ≡ α Cen C. Remarkably, it is not even clear at the present time whether the standard Newtonian theory of gravity, the great stalwart underpinning of astronomical dynamics, even applies to stellar systems such as α Cen AB and Proxima. This is one of the deeper modern-day mysteries that this book will explore in later pages.

Proxima Hiding in the Shadows

Proxima, again, for all of its adjacency to the Sun, is far from being an obvious star. It cannot in fact be seen by the unaided human eye, and indeed a relatively large-aperture telescope is required to reveal its meager light. It is because of this low intrinsic brightness that Proxima's very existence and nearest stellar neighbor status was only established in the early twentieth century. Remarkably, as will be seen, Proxima as a red dwarf star belongs to the most populous class of stellar objects within our Milky Way Galaxy; for every Sunlike star in the galaxy, there are eight to ten Proxima-like stars. And yet, the unaided human eye can see not one such representative of this vast indigenous population. Adventure, exploration, and discovery not only open our collective eyes to the greater Universe, they also take us beyond our direct human senses, enabling us to *see* those places where likely only the mind will ever go.

As we encounter the centennial of Proxima's discovery, it seems only appropriate to consider how our understanding of the α Centauri system has changed and how astronomical knowledge has evolved during the past 100 years. Indeed, since the discovery of Proxima our appreciation of the stars and planets and the greater cosmos has changed almost beyond recognition. When Proxima was first identified in 1915, Einstein's general theory of relativity, one of the great cornerstones of modern physics, was still a year away from publication. The Bohr model, the first quantum mechanical description for the workings of the atom, was barely 2 years old. Hubble's law and the discovery of the expanding universe were still 15 years in the future. The first public TV broadcast was likewise 15 years distant, and the radio signal bubble centered on Earth was barely 10 light-years across. Indeed, the feeble radio waves representing the very first public broadcast transmitted from the Metropolitan Opera House in New York on January 13, 1910, had only swept past α Centauri the year before Proxima was first identified. Today, over 100 years later, Earth's radio bubble encompasses a volume containing well over a thousand stars.

The α Centauri system became the Sun's closest stellar neighbor about 50,000 years ago. Since that time, it has watched over the rise of human history and the development of civilization as we know it; Proxima in turn, since its discovery, has overseen the incredible advancements in the technologies that define our modern computer-driven and hyperlinked society. The stars of α Centauri will remain our closest stellar companions for another 72,000 years, and we may but dream what continued changes will take place on Earth during this extended period of time. But for all this, as α Centauri drifts ever further away from the Sun, dropping below the threshold of naked-eye visibility in about one million years from the present, its story is far from over – as will be seen in the main text. Indeed, the story of Proxima will be played out within the confines of our evolving galaxy over the next many trillions of years, by which time the Sun and α Centauri A and B will have long cooled off to degenerate black dwarfs. Who knows where humanity might be such colossal timescales hence? What is certain, however, is that we will have changed beyond all present-day recognition and cognition, but then, not just ourselves, the entire observable universe will be very different when Proxima Centauri dies.

Some Notes on Units and Nomenclature

Astronomy texts and astronomers are notoriously bad at mixing their units, a result mostly due to a long history and the sheer scale of the subject. In general, the units to be used in this text will be those of the System International, with distances expressed in meters and masses expressed in kilograms. Other units, however, will be used when planetary and stellar distances are being considered.

It is often said that unit changes are done in order to avoid writing down large numbers, but this of course is just psychological camouflage; the numbers, no matter what the units, measure the same thing. For all this, however, we shall encounter the astronomical unit, the parsec, and the light-year. The first two of these new units follow naturally from the size of Earth's orbit about the Sun (corresponding to 1 astronomical unit, or au) and the distance to a star for which the half-annual parallax is 1 arc sec (corresponding to 1 parsec, or pc). The third distance is derived from the constancy of the speed of light 2.99792×10^8 m/s and the number of seconds in an average Gregorian year, with 1 light-year = 0.3066 pc = 63,239.8 au = 9.4605×10^{15} m. Angles will normally be expressed in degrees or in the subunits of arc minutes (1/60th of a degree) and arc seconds (1/60th of an arc minute). On occasion, the unit of milliarc seconds (mas) will appear, with 1 mas = 1/1,000th of an arc second. On a very few occasions, the angular unit of radians will be introduced, with 2π radians = 360°.

Units for stellar mass, luminosity, and radius will typically be expressed in solar units, with 1 $M_\odot$ = 1.9891×10^{30} kg, 1 $L_\odot$ = 3.85×10 26 W, and 1 $R_\odot$ = 6.96265×10^8 m. The Sun unit will be explicitly implied through the use of the symbol $\odot$. Temperatures will be expressed in Kelvin, with the zero Kelvin mark corresponding to the absolute zero point of temperature. The convention, for various historical reasons, is also to write just Kelvin rather than degrees Kelvin. In terms of the more familiar everyday temperature scales, 0 K = –273.15 °C = –459.67 °F.

Several methods will be used to identify individual stars within the text. In some cases, a star has a historical name such as Sirius (derived from the Greek word for "scorching"), which is the brightest star (next to the Sun, of course) observable to the unaided human eye at the present epoch. Sirius is also the brightest star in the constellation of Canis Major (The Great Dog), and its Bayer identification[5] is accordingly α Canis Majoris. Murzim, the second brightest star in Canis Major, is identified as β Canis Majoris and so on through the Greek alphabet for the remaining principle stars in the constellation. Stars can also be identified through their various catalog numbers, and accordingly Sirius in the Henry Draper catalog of stars is identified as HD 48915. In the Hipparchos data catalog, Sirius is identified as HIP 32349. Most of the time, this extended range of celestial monikers – Sirius has at least 58 aliases – is not something for us to worry about, but it is worth being aware of the fact that different names and identification numbers do exist for essentially all cataloged stars.

The identification scheme for stars within a binary system is mostly self-evident, and we have already used it above, but for completeness the two components in a double star system are labeled A and B, with the A label being applied to the more luminous component. Sirius, once again for example, is actually a binary system, and the star that we see with our eyes should technically (at least in the modern era) be called Sirius A. Its small, low-luminosity white dwarf companion, Sirius B, is only observable in a relatively large-aperture telescope, and it was not actually observed until 1862, when Alvin Clark first tested his newly constructed telescope incorporating an 18-in. (0.457 m) objective lens. Sirius A, of course, was observed and known about since before recorded history. It is sometimes convenient to explicitly identify a star as being a binary system, and accordingly Sirius might be described as the system Sirius AB. Likewise, α Centauri can be described as α Centauri AB and more simply still as α Cen AB.

With the discovery by Michel Mayor and Didier Queloz in 1995 of the first exoplanet in orbit around the Sunlike star 51 Pegasi, astronomers needed a new nomenclature scheme to identify nonstellar components. Although there is as yet no officially sanctioned scheme, the most commonly used method identifies the various planets within a specific system with a lowercase letter starting with the letter b and then working systematically through the alphabet. The planet identification label starts with the letter b since technically according to the scheme, the parent

[5] German astronomer Johann Bayer introduced this scheme in his 1603 *Uranometrica* star atlas.

star corresponds to system subcomponent a. Astronomers, however, generally ignore this latter convention and drop the "a" label for the star. (Common usage and historical precedent will always triumph over any set of conventions whether officially sanctioned or not.)

So with this entire preamble in place, the planet discovered by Mayor and Queloz is identified as Pegasi b. Just to make life a little more complicated, planet 51 Pegasi b is sometimes unofficially referred to as Bellerophon after the mythological Greek hero who tamed the winged horse Pegasus. If a second planet were to be found to orbit 51 Peagasi = 51 Pegasi a, it would be identified as 51 Peagasi c.

The planet-labeling sequence is based upon the time of discovery rather than orbital distance from the parent star, and accordingly planet b need not, for example, be the innermost planet within a multiple-planet system. For planets within binary star systems, both the star component and the planet need to be specified. So, for example, if a planet were to be found in orbit about Sirius A = Sirius Aa, it would be identified as Sirius Ab = α Canis Majoris Ab. As we shall see later on, the first planet to be detected in the α Centauri AB system is in orbit around α Cen B = α Cen Ba, and accordingly it is identified as α Cen Bb.

"It glows above our mighty sea-laved isle,
Changing and flick'ring in the arch of God,
Where miles and miles of grassy levels smile,
And where the unsung pioneers have trod.
Alpha Centauri! See the double star
That gleams as one above the smoke-drift cloud,
Above the groves of Redwood and Bethar,
Or where the checked Pacific thunders loud.
Star of my home! When I was a child,
Watching, and fearful of the coming years,
You bade me learn the story of the wild,
You bade me sing it low to stranger ears!"

– From the poem "Alpha Centauri" by Mabel Forrest, (1909)

Contents

About the Author

Martin Beech is a full Professor of astronomy at Campion College at the University of Regina and has been teaching courses relating to planetary science, stellar structure, and the history of astronomy for nearly 20 years. His main research interests are in the area of small Solar System bodies (asteroids, comets, meteoroids, and meteorites) along with the development of computer models relating to the structure and evolution of stars. Minor planet (12343) Martin Beech has been named in recognition of his contributions to meteor physics. He has written several books with Springer previously and published many articles in science journals.

1. Discovery, Dynamics, Distance and Place

1.1 First Light

It was a clear and windless winter's evening in early July when this author first saw α Centauri. Brought into sharp focus by a telescope at the Stardome Observatory Planetarium in Auckland, New Zealand, its light was of a cold-silver. The image was crisp and clear, a hard diamond against the coal-black sky. The view was both thrilling and surprising. To the eye α Centauri appears as a single star – the brightest of 'the pointers.' Indeed, to the eye it is the third brightest 'star' in the entire sky, being outshone only by Sirius and Canopus (see Appendix 1).

Through even a low-power telescope, however, a remarkable transformation takes place, and α Centauri splits into two: it is a binary system. Composed of two Sun-like stars, α Cen A and α Cen B orbit their common center once every 80 years, coming as close as 11.3 AU at periastron, while stretching to some 35.7 au apart at their greatest separation (apastron). Perhaps once in a human lifetime the two stars of α Centauri complete their rounds, and they have dutifully done so for the past six billion years (as we shall see later on). Having now completed some 75 million orbits around each other, the two stars formed and began their celestial dance more than a billion years before our Sun and Solar System even existed. The entire compass of human history to date has occupied a mere 125 revolutions of α Cen B about α Cen A in the sky, and yet their journey and outlook is still far from complete. Indeed, the α Centauri system will outlive life on Earth and the eventual heat-death demise of the planets within the inner Solar System.

However, a hidden treasure attends the twin jewels of α Centauri. A third star, Proxima Centauri, the closest star to the Sun at the present epoch, lurks unidentified in the background star field.

M. Beech, *Alpha Centauri: Unveiling the Secrets of Our Nearest Stellar Neighbor*, Astronomers' Universe, DOI 10.1007/978-3-319-09372-7_1,

Proxima is altogether a different star from either α Cen A or α Cen B. It is very much fainter, smaller in size and much less massive than its two companions, and it is because of these diminutive properties that we cannot see it with the unaided eye. These are also the reasons why it will survive, as a *bona fide* star, for another five trillion years. Not only will Proxima outlive humanity, the Sun and our Solar System, it will also bear witness to a changing galaxy and observable universe.

For all this future yet to be realized, however, the story of Proxima as written by human hands begins barely a century ago, starting with its discovery by Robert Innes in 1915 – at a time when civilization was tearing itself apart during the first Great War.

However, we are now getting well ahead of ourselves. Let us backtrack from the present-day and see what our ancestors made of the single naked-eye star now called α Centauri.

1.2 In Honor of Chiron

The constellation of Centaurus is one of the originals. It has looked down upon Earth since the very first moments of recorded astronomical history. It is the ninth largest, with respect to area in the sky, of the 88 officially recognized constellations, and it was described in some detail by Claudius Ptolemy in his great astronomical compendium written in the second century A.D. Ptolemy placed 37 stars within the body of Centaurus, but modern catalogs indicate that there are 281 stars visible to the naked eye within its designated boundary (Fig. 1.1). The two brightest stars, α and β Centauri, however, far outshine their companions, and they direct the eye, like a pointillist arrow, to the diminutive but iconic constellation of Crux – the Southern Cross.

In order to ease the discussion that is to follow let us, with due reverence, refine and reduce the skillfully crafted map of Fig. 1.1. Removing the background clutter of faint stars, and minimizing still further the constellations markers, we end up with just ten stars. These stars, our minimalist centaur, are shown in Fig. 1.2. Even without our pairing down, an abstract artist's eye is required to unravel the hybrid body being traced out, point by point, by the stars in Centaurus. This twisted perception is perhaps

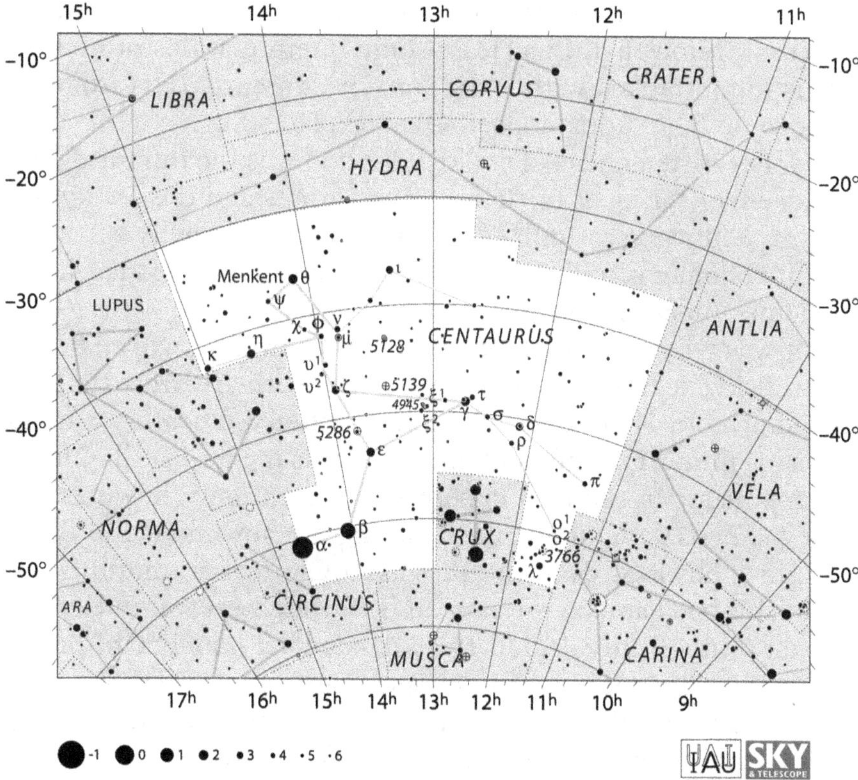

FIG. 1.1 The constellation of Centaurus (Image courtesy of the IAU and Sky & Telescope, Roger Sinnott and Rick Fienberg, Centaurus_IAU.svg)

even more compounded by the fact that the centaur, so revealed by the stars, is a mythical beast, created entirely by the human imagination rather than the level-headed workings of natural selection and evolution. The half-man, half-horse centaurs take us back to a time in history that was ancient even to the ancient Greeks; to a time when capricious gods were thought to play out their political games, jealousies and in-fighting on Earth. The centaur, in literature at least, has typically been thought of as being fierce, when and if the need arises, but generally learned and wise. C. S. Lewis in his *Chronicles of Narnia* portrays them as noble creatures that are slow to anger but dangerous when inflamed by injustice. J. K. Rowling, in her *Harry Potter* series of books, places the centaurs in the Forbidden Forrest close by Hogwarts, making them both secretive and cautious. For all this, however, they are taken as being wise and skilled in archery.

Chiron is generally taken to have been the wisest of centaurs, and it is Chiron that, in at least some interpretations of mythology, is immortalized within the stars of Centaurus. His story is a tragic one. The Roman poet Ovid explains in his *Fasti* (The Festivals, written circa A.D. 8) that Chiron was the immortal (and forbidden) offspring of the Titan King Cronus and the sea nymph Philyra. Following a troubled youth Chiron eventually settled at Mount Pelion in central Greece and became a renowned teacher of medicine, music and hunting.

It was while teaching Heracles that Chiron's tragic end came about. Being accidentally shot in the foot with an arrow that had been dipped in Hydra's blood, Chiron suffered a deathly wound, but being immortal could not die. For all his medical skills Chiron was doomed to live in pain in perpetuity. Eventually, however, the great god Zeus took pity on the suffering centaur, and while allowing him a physical death he preserved Chiron's immortality by placing his body among the stars. It is the brightest star in Centaurus that symbolically depicts the wounded left hoof of Chiron.

In the wonderful reverse sense of reality aping mythology the flight of Chiron's death-bringing arrow is reenacted each year by the α Centaurid meteor shower. Active from late-January to mid-February the α Centaurid shooting stars appear to radiate away from Chiron's poisoned hoof (see Fig. 1.2). Bright and swift, the α Centaurid meteors are rarely abundant in numbers, typically producing at maximum no more than 10 shooting stars per hour. In 1980, however, the shower was observed to undergo a dramatic outburst of activity. A flurry of meteors were observed on the night of February 7, with the hourly rate at maximum rising to some 100 meteors, a high ten times greater than the normal hourly rate. The small grain-sized meteoroids responsible for producing the α Centaurid meteors were released into space through the outgassing of a cometary nucleus as it approached and then rounded the Sun, but the orbit and identity of the parent comet is unknown. It is not presently known when or indeed if the α Centaurid meteor shower will undergo another such outburst. Only time, luck and circumstance will unravel the workings of this symbolic, albeit entirely natural, annihilation re-enactment.

Arabic astronomers during the first millenium A.D. knew α Centauri as *Al Riji al Kentauris,* which translates to

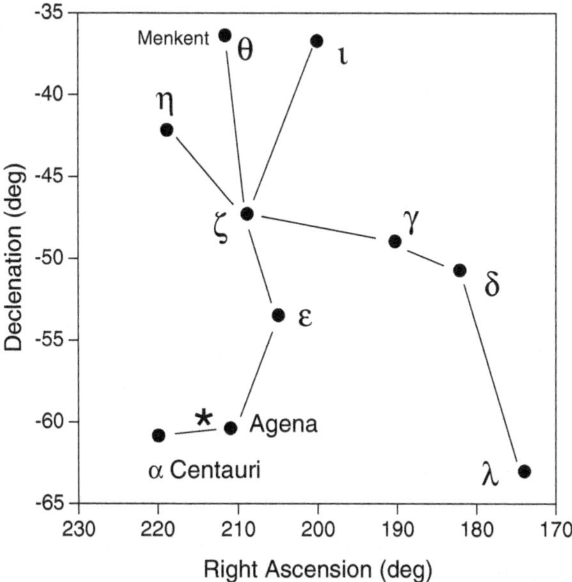

FIG. 1.2 A minimalist star map for Centaurus. The ten stars depicted are all brighter than magnitude +3.2 and easily visible to the naked eye. The radiant location for the α Centaurid meteor shower, at about the time of its maximum activity, is shown by the * symbol

"the Centaurs foot," and it is from the Latinized version of this expression that we obtain Rigel Kentaurus.

Strangely astronomers have never really warmed to any of the names historically given to α Centauri, and to this day it officially has no specified moniker. Nicolaus Copernicus in his epoch-changing *De Revolutionibus* (published in 1543) reproduced Ptolemy's star catalog almost verbatim, and there the brightest star in Centaurus is simply described as the one "on top of the right fore-foot" – a description that hardly inspires distinction.[1] A search through the SIMBAD[2] database maintained at the University of Strasbourg reveals a total of 33 identifiers for α Centauri, ranging from the rather dull FK5 538 to the extensive (but still dull) J143948.42-605021.66, but no common name is presented.

[1] Ptolemy refers to the right foot, since his imagined view is that of a god-like observer looking down on the sphere of the heavens. For us mortals, on Earth, α Centauri appears as the left front foot.

[2] SIMBAD = Set of Identifications, Measurements, and Bibliography for Astronomical Data.

In addition to the colloquial Rigel Kent, α Centauri is also known as Toliman. This later name is obscure, but it has been suggested that it refers to the root or offshoot of a vine and is reflective of the literary notion that centaurs would often carry a vine-entwined staff. Other authorities have suggested that Toliman is derived from the Arabic *Al Zulman* meaning "the ostriches," although no specific reason is given for this avian association.

The historical naming confusion over α Centauri is further echoed by its companion, and the second brightest star in the constellation, β Centauri. This star is variously known as Hadar and/or Agena. The word Hadar is derived from the Arabic for "ground" or "soil," while Agena is derived from the Latin words for the knee. The third brightest star in the constellation θ Centauri is known as Menkent, which is derived from the Arabic meaning "shoulder of the centaur," although this being said, Menkent is sometimes depicted as indicating the location of the head of the cenataur.

Chiron is not the only centaur that adorns the sky. Indeed, he has a doppelganger in the constellation of Sagittarius. Also a Southern Hemisphere constellation, Sagittarius is the Archer who is carefully aiming his celestial arrow at the menacing heart of Scorpio – the celestial arthropod. Some classical authorities have linked the story of Chiron to Sagittarius, rather than the constellation of Centaurus, while others claim that Chiron invented the constellation of Sagittarius (in his own image?) to guide the Argonauts in their quest for the Golden Fleece. Irrespective of where Sagittarius fits into the mythological pantheon, it is clear that he guards the galactic center, his imagined arrow pointing almost directly towards the massive black hole (identified with the strong radio source Sagittarius A*) located at the central hub of the Milky Way's galactic disk.

1.3 *Te taura o te waka o Tama-rereti*

To the aboriginal Maoris of New Zealand the sky is alive with symbolism and mythology. Their ancestors were no less imaginative than the ancient Greeks. The Maori sky is also a vast seasonal clock and navigational aid, with the helical rising of Matariki (the asterism of the Pleiades) near the time of the mid-winter solstice, setting the beginning of each new year. At this moment the largest of the Maori constellations *Te Waka o Tame-rereti* (The Great Waka[3]) stretches right across the southern horizon – arching some 160° around the sky. The Great Waka is made-up of the Milky Way and its associated brighter stars. The prow of the Waka is delineated by the curve of stars in the tail of Scorpio, and its anchor (*te punga*) is symbolized by the constellation of the Southern Cross (Crux) – which at the time of the Maori New Year is located low in the sky and due south. Connecting *te punga* to *Te Waka o Tame-rereti* was the anchor line (*Te taura o te waka o Tama-rereti*), and two of the bright links in the anchor chain were α and β Centauri. With the Great Waka so anchored and riding the southern night sky at the time of the New Year, the important seasonal and navigation stars are also displayed. To the west of *te punga* is *Rehua* (Antares = α Scorpio), to the east is *Takurua* (Sirius = α Canis Majoris). Above the Great Waka is *Atutahi* (Canopus = α Carinae).

The placing of the stars in the Maori creation cycle is associated with the voyage of *Tama-rereti*, who was charged to bring light into the world and make a great cloak for *Rangi* – the personification and essence of things made. Rangi's cloak is depicted by *Te Ikaroa* (the Milky Way), which was made from the lesser stars spilling out of the Great Waka.

To all cultures, not just the Maoris, the night sky is a vast storyboard. It tells the time, the seasons and guides the explorer, and it also displays an ancient echo of the deeper mysteries pertaining to the act of creation and the workings of elemental forces and nurturing gods. Although α Centauri is not one of the culturally important stars of the Maori (indeed, there is no specific name for it), the fact that it helps anchor the Great Waka to the sky, enabling thereby both heavenly permanence and predictability, makes it a star of metaphorical strength and stability. For the Maori α Centauri anchors the great ship of migration to the sky. It is also the embodiment of place in Mabel Forrest's poem – as reproduced in the introduction. Indeed, for Forrest α Centauri defines the very essence of what might reasonably be called southern-ness,

[3] A waka is a long, narrow-beamed canoe.

and such feelings provide us with a new perspective. It is the other Janus-face of α Centauri that we see now. It is the star that pulls us to the heavens, engendering dreams of interstellar travel, and it is the star that fixes location and domicile.

1.4 And in Third Place...

Coming in third is no bad thing if one is competing in a sporting competition, but for α Centauri, being the third brightest 'star' discernable to the naked eye has largely resulted in its being written out of cultural history – and this, in spite of its embodiment of the southern *genius loci*. Sirius and Canopus are the first and second brightest stars visible to the unaided human eye,[4] and each of these heavenly lamps has, at one time or another, been subject to deep religious and cultural veneration.

To the ancient Egyptians, Sirius was associated with Isis, the goddess of motherhood, magic and fertility, and its helical rising each July was seen as a sign for farmers to prepare for the Nile inundation – that vital, life-sustaining, annual flood that would ensure the successful growth of the next crop. Again, to the ancient Egyptians Canopus, visible for just a few months of the year, located low on the southern horizon, became known as an important marker star, being associated with both physical navigation (showing the southern direction) and the spiritual journey of dead and departed souls.

Although it appears that no deep spiritual associations have been attached to α Centauri its distinctive nearness to β Centauri (Agena) in the sky has not gone unnoticed. To the ancient Inca society of South America, the two stars were the eyes of the mother llama; to the Australian aborigines, the two stars signified the story of the hunted emus and frightened possum. In Chinese astronomical lore, however, α Centauri is located in the asterism of the Southern Gate (associated with the Horn mansion within the Azure Dragon of the East), and it is simply the fifth star in the pillars of the library house.

---

[4] In terms of apparent magnitude ranking (see Appendix 1 in this book), α Cen A is the third brightest star in the sky, with $m = -0.27$, α Cen B is the 21st brightest star with $m = +1.33$, while β Centauri (Agena) is the 10th brightest star, with $m = +0.60$.

Even in modern times being third brightest star has counted against α Centauri. This is perhaps best exemplified in the Brazilian national flag, arguably the most detailed astronomical flag ever produced. Blazoned across the central circle of the flag are the words *Order e Progresso*, words inspired by French philosopher Auguste Comte's order and progress credo of positivism. Additionally, within the central circle are shown 27 stars symbolizing the Brazilian State and its federal districts. The stars are projected onto the flag as they would appear to an imagined external observer (that is, one looking down upon Rio de Janeiro from outside of the heavenly vault) at 08:30 on November 15, 1889 – the moment of Brazil's independence from Portugal. The 27 stars nicely pick out the locations of Sirius and Canopus, and they delineate the constellations of the Southern Cross, Scorpius, Hydra and even Triangulum Australe, but α Centauri is nowhere to be seen. Indeed, no stars from Centaurus are depicted upon the flag at all. Likewise, the national flags of New Zealand and Australia have adopted the stars of the Southern Cross as their distinctive and identifying feature[5] – α Centauri and The Pointers relegated to apparent insignificance.

1.5 Over the Horizon

To the author, who lives in the prairies of central Canada, α Centauri is sadly never observable; it literally never rises above the horizon. How far south from Canada, therefore, must one travel in order to catch a first glimpse of our Centurian stellar quarry?

The answer is in fact quite straightforward to obtain and is just a matter of geometry and angle determination. Astronomers fix the position of a star in the sky according to its right ascension (RA) and angle of declination (δ). These two coordinates are similar to those in an ordinary Cartesian X-Y graph, as seen, for example, in the financial section of any newspaper on any day of the year, but they are specialized to describe an imagined spherical

[5] At first glance it might appear that α Centauri is featured on the Australian flag, but the large seven-pointed star under the defiled union jack symbolizes the federation of the seven Australian states.

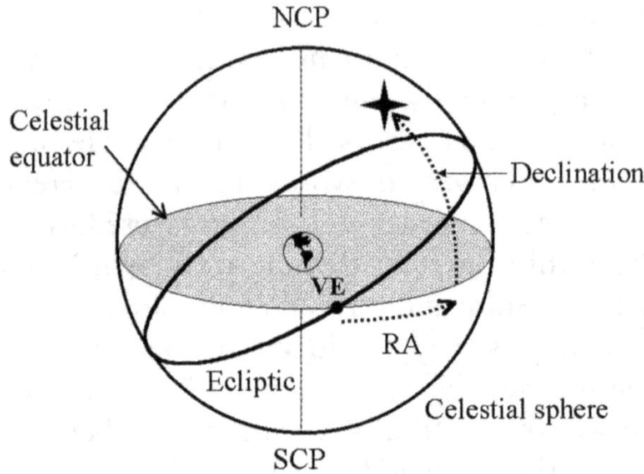

FIG. 1.3 The celestial sphere and the astronomical coordinate system. The celestial equator corresponds to the great circle projection around the sky of Earth's equator. The ecliptic corresponds to the location of the Sun, with respect to the background stars as seen from Earth, during the course of 1 year

sky, the celestial sphere, with a specific origin, set according to the sky intercept position of the ecliptic and celestial equator (Fig. 1.3) – the so-called first point of Aries.[6] The sky coordinates of α Centauri in 2000 can be taken from any standard table of star positions, and accordingly: RA = 219.90° and δ = –60.83° – the negative sign indicates that α Centauri is located south of the celestial equator. The essential geometry of the situation in question is illustrated in Fig. 1.4. Fortunately we need only know the declination of α Centauri, to answer the question at hand, and this explains why the figure can be drawn in just two dimensions rather than three. The right ascension coordinate principally determines the angle of α Centauri around the sky for a given observer.

From Fig. 1.4, the angle that α Centauri subtends to the celestial equator, which is simply the projection of Earth's equator onto the celestial sphere, is δ degrees. An observer S, located at a latitude of λ = δ, will be able to see α Centauri directly overhead. With this information in place, the location for an observer N where α

[6] Somewhat confusingly, the location for the origin point for right ascension is no longer in the constellation of Aries; rather it is now located in the constellation of Pisces.

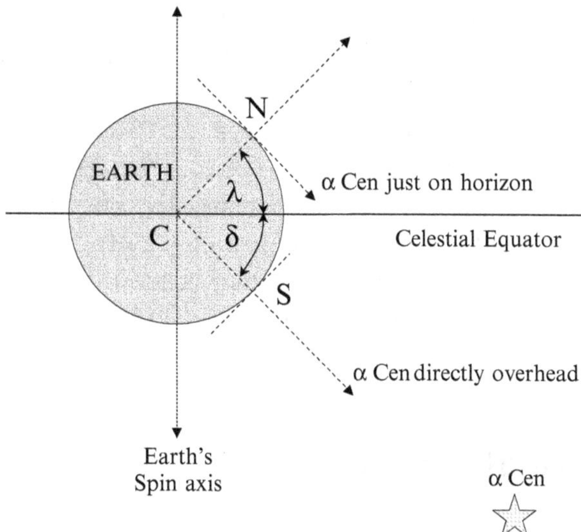

FIG. 1.4 The visibility condition for α Centauri

Centauri will just peak above the horizon, satisfying the just visible condition, can be determined. Accordingly, the latitude of observer N will be 90° north of observer S. Given that $\delta = -60.83°$, the latitude at which α Centauri will begin to just peak above the horizon will be $\lambda = \delta + 90 = 29.17°$. For the author, therefore, located at latitude 50.45° north of the equator, a journey encompassing some 50.45–29.17 = 21.28° of latitude due south will be required before a glimpse of α Centauri could be made. This travel requirement would place the author not too far away from Guadalajara, Mexico.

In contrast to the horizon 'peaking' condition, we can also determine a second, special observability condition for α Centauri; specifically, the latitude on Earth below which, that is, south of, it will never set below the horizon. This so-called circumpolar condition is directly related to the angular distance of α Centauri away from the south celestial pole (SCP). (See Fig. 1.3 and Sect. 1.6 below.). Since, by definition, the SCP has a declination of −90°, so the angle between the SCP and α Centauri is: $-90-(-60.83) = -29.17°$.

What this now tells us is that, once the altitude of the SCP is 29.17° or more above an observer's horizon, so α Centauri, as it completes one 360° rotation around the SCP during the course of

1 day, will never drop below the horizon. Accordingly, α Centauri will be a circumpolar star for all locations south of −29.17° latitude. This region encompasses all of New Zealand, the southern half of Australia, the tip of South Africa, and South America below about the latitude of Santiago in Chile. The only landmass on Earth where α Centauri might potentially be observed by the unaided human eye during a complete 24-h time interval is Antarctica – and in this case the viewing would need to be made during the time of complete darkness associated with the Antarctic winter.

1.6 Practical Viewing

Our distant ancestors not only used the heavens as a vast clock, ticking off the hours, days and seasons according to the visible stars and constellations, they also used the sky for navigation. Once the concept of the celestial sphere had been established, it was evident that there were two special points on the sky about which all the stars appear to rotate. These special points, called the celestial poles, are simply the projection of Earth's spin axis onto the celestial sphere. There is accordingly a north and a south celestial pole.

Northern observers have been fortunate during the last few thousand years to have a reasonably prominent constellation (Ursae Minoris) to guide the eye and a reasonably bright star (α Ursae Minoris = Polaris) to indicate the position of the north celestial pole (NCP). Find Polaris and you instantly know where north is, and just as importantly this guide star works on any night of the year – as Shakespeare so poignantly reminds us in Sonnet 116, "It [Polaris] is an ever-fixed mark that looks on tempests and is never shaken; it is the star to every wondering bark."[7]

Navigators of the southern oceans have been less fortunate than their northern hemisphere cousins. In principle, once a journey has taken a navigator south of the equator the south celestial pole can be used in the same way as the north celestial pole – that is, it can be used to find the direction of due south. In practice,

[7] A bark is a three-masted, square-rigged, ship.

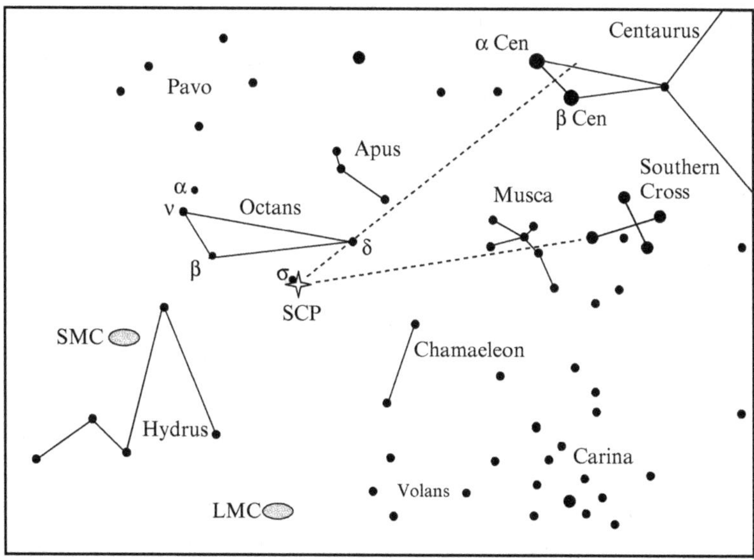

FIG. 1.5 Using The Pointers (α Centauri and β Centauri) and the Southern Cross to determine the location of the south celestial pole (SCP). The south celestial pole is indicated by the cross close to the star σ Octantis. The Large (LMC) and Small (SMC) Magellanic Clouds, visible to the human eye as faint smudges on the sky, are satellite companions to the Milky Way Galaxy

however, it is far from easy to identify the location of the south celestial pole, since it does not coincide with any bright star or prominent constellation. The south celestial pole is appropriately enough located in the constellation of Octans, named after the octant (one-eighth of a circle) navigational instrument, and technically the closest marker to the pole is the just visible to the naked-eye star σ Octantis.

None of this is particularly helpful, however, as a practical guide, and navigators have long used a less precise but much easier to apply method for finding the south celestial pole. The trick is to use The Pointers, made-up of α Centauri and β Centauri, and the Southern Cross (Crux). Figure 1.5 shows a star map of the region surrounding the south celestial pole. In order to find the pole an observer must construct two imaginary lines (the dashed lines in Fig. 1.5), one leading through the longer staff of the Southern Cross and the other at right angles to the point midway between The Pointers. Where these lines intercept on the sky is the approximate location of the south celestial pole. By dropping a line directly downward from the south celestial pole the direction of due south will be identified on the horizon.

1.7 Slow Change

Although the relative distances between the stars appear fixed over time, they do, in fact, undergo a slow and steady independent motion. Not only, in fact, do the viewing conditions for seeing a constellation above a specific observer's horizon change over the centuries, but so too does the spacing between the stars in the constellation. The first motion relates to the changing orientation of Earth's spin axis, while the second motion relates to the spatial movement of the stars themselves. The stars are indeed free spirits. Shakespeare was only partly right when he described Polaris (α Ursae Minoris) as being a fixed point in the sky – that is, located at the north celestial pole.

Polaris and the NCP are presently nearly coincident, but they were not so in the distant past and they will not be so again in the distant future. Due to the precession of Earth's spin axis – an effect produced by the non-symmetric mass distribution of Earth and the gravitational influence of the Sun and Moon – the location of the NCP, with respect to the background stars, traces out a large (23.5° radius) circle on the sky. It takes the NCP some 26,000 years to complete one precession-induced cycle through the heavens, a motion that amounts to about one degree in the sky per good human lifetime of 72 years. This precession cycle causes the gradual drift of the constellations with respect to the celestial coordinate system (Fig. 1.3), and as a result of this Centaurus has slowly been tracking southward, with respect to the celestial equator, over the past many millennia. When first placed in the heavens, at the dawn of human history, the constellation of Centaurus was situated much closer to celestial equator when it was then delineated, and it would have been visible throughout much of the northern hemisphere – the region, in fact, from which it is now mostly excluded from view. As future millennia pass by, however, Centaurus will once again move closer to the celestial equator, but by then its star grouping will have begun to change beyond present-day recognition.

In reflex sympathy with the movement of the NCP, so the south celestial pole also moves around its own circular path with respect to the stars. Just as Polaris will eventually turn out to be a false guide, no longer leading voyagers northward, so The Pointers and the Southern Cross will eventually fail to locate the SCP. This change will come about only slowly, by human standards, and were it not for the individual motions of the stars our stellar sign-posts would correctly pick out the SCP every 26,000 years. For The Pointers, however, this epoch is the only moment in the entire history of the universe (literally the universe past and the one yet to come) when they will act as trustworthy guides. The reason for this relates to the rather hasty manner in which α Centauri moves across the sky. The speed with which α Centauri is moving through space, relative to the Sun (for details see Appendix 2), is about 22.5 km/s, and as seen from Earth this translates into a proper motion of some 3.7 arc sec per year across the sky with respect to the much more distant "fixed"[8] stars.

The proper motion of α Centauri is the 12th largest recorded, and it comes about largely through its present close proximity to the Sun rather than the result of an exceptional space velocity.[9] The index finger held out at arm's length covers an angular arc of about 1° in the sky, and it will take α Centauri some 973 years to accumulate this same angular shift. Remarkably, therefore, in 2000 B.C., when the ancient Babylonian sky watchers first began to name the heavens, α Centauri was located some 4° (four fingers' width) away from its present location with respect to Agena and the other stars in Centaurus. The centaur had a much more aggressive foreleg stance in the distant past.

The proper motion of a star is dependent upon its distance from the Sun as well as its actual space velocity (for details see Appendix 2). We will discuss how the distances to the stars are measured further below, but suffice to say at present, the stars that constitute the constellation of Centaurus range in distance from

---

[8] The distant stars are not actually fixed in space, of course. Rather, their vast distance away from us means that the time required to accumulate any measurable shift in the sky is determined on a timescale of many millennia.

[9] Barnard's Star has the largest known proper motion, moving across the sky at a rate of 2.78 times faster than α Centauri. Of the stars within 5 pc of the Sun, Kaptenyn's Star has the highest space velocity of 293 km/s.

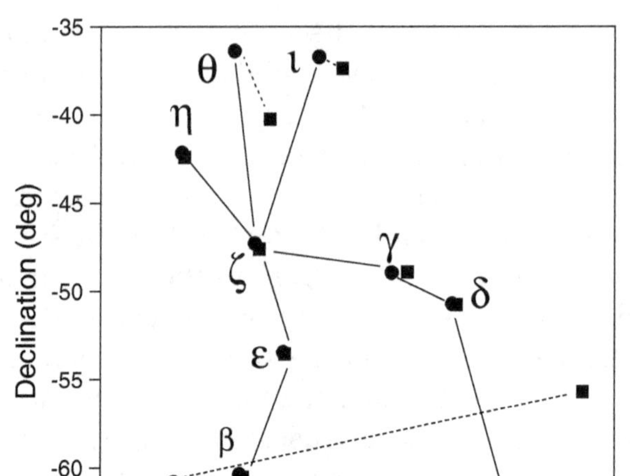

FIG. 1.6 The proper motion change in the relative positions of the principle stars of Centaurus during the next precession cycle of Earth. *Filled circles* indicate positions at the present time, while *filled squares* indicate locations 26,000 years hence

4.3 light years (for α Centauri) to some 427.3 light years away (for ε Centauri – see Fig. 1.2). Given these distances, the range in proper motion is therefore also quite large, varying from 3.7 arc sec per year (for α Centauri) to a lowly 0.02 arc sec per year (again, for ε Centauri).

Because of this difference in proper motion characteristics the stars in Centaurus will gradually shift relative to each other, and it is, in fact, for this reason that The Pointers (α and β Centauri) will ultimately act as false guides to the south celestial pole (Fig. 1.5). Figure 1.6 shows the accumulated shift in the relative positions of the principle stars in Centaurus over the next 26,000 years – i.e., the time corresponding to one complete precession cycle of Earth. Clearly, α Centauri is the high flyer in this time interval, and by the end of the next precession cycle it will occupy a position consistent with the delineation of the centaur's tail rather than its present hoof.

After α Centauri, the most rapid proper motion movers within the constellation are θ Centauri (Menkent) and ι Centauri.

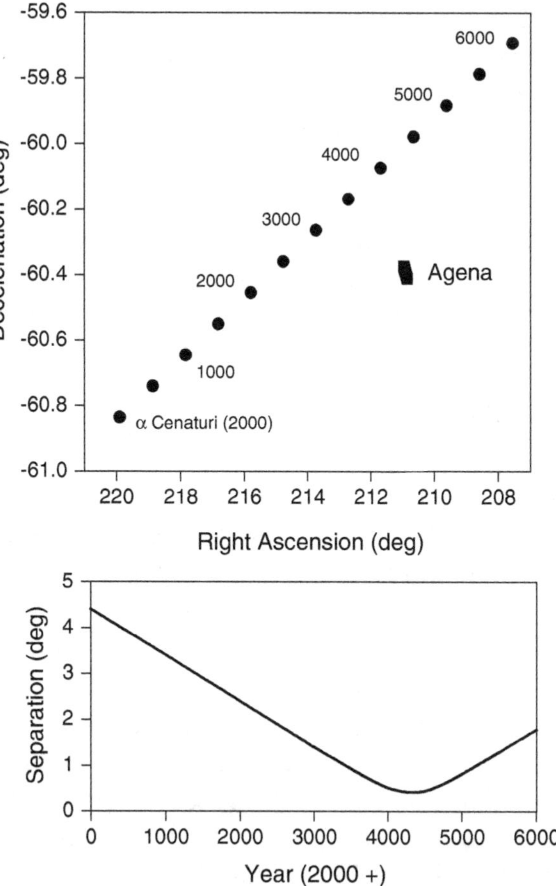

FIG. 1.7 The proper motion (*upper* diagram) and relative separation (*lower* diagram) of α and β Centauri. The proper motion of Agena (β Centauri) is some 925 times smaller than that derived for α Centauri. The numbers indicate the year A.D. with "α Centauri – now" corresponding to the year 2000

Of the other stars, their accumulated proper motion is very small, and their relative positions hardly change. The body and back legs of the centaur are barely going to twitch during the next 26 millennia.

Agena (β Centauri) is located some 89 times further away from the Sun than α Centauri, and it accordingly has a small proper motion of just 0.04 arc sec per year. Figure 1.6 shows that the proper motion of α Centauri is carrying it rapidly towards and then away from Agena, and this motion will result in an interesting stellar conjunction in about 4,000 years hence. Figure 1.7 shows the proper motion positions for α Centauri and Agena over the next 6,000 years in detail. In this time interval Agena hardly moves at all, while α Centauri gallops across more than 10° of the sky. The time of closest approach will occur about the year A.D. 6400, and at that time the two stars will be a little less than half a degree apart in the sky – as opposed to their present 4.4° separation. This future close pairing of two of the brightest stars in the sky will be a jewel of a stellar spectacle to see, but sadly not a view for any current reader to behold.

1.8 The Splitting of α Centauri

Comets have historically been cast as the harbingers of doom, their diaphanous tails casting a foreboding arc across the sky for both lowly plebian and king alike to see. For α Centauri, however, the comet of 1689 was not so much a messenger of despair but a vector of revelation. It was by following the course of the new comet through the sky on the night of December 21 that the be-telescoped gaze of Jesuit missionary Jean Richaud, working from Pondicherry in India, was brought to bear upon the principle star of Centaurus. To Richaud's great surprise and presumed delight it was thereby transformed from the single star, as perceived by the naked eye, into a double or binary star; "The two stars seemed to be practically touching each other," he wrote to a friend. The true double nature of α Centauri, otherwise conjoined by low resolution and the shear withering of distance, was revealed for the first time in human history.

In recognition of Richaud's discovery one of the official identifiers for α Centauri is RHD 1, with the 1 being somewhat optimistic, since no other stars have an RHD identifier. Writing almost 150 years after Richaud's discovery, the great John F. W. Herschel, observing from Johannesburg in South Africa described the α Centauri pairing as being, "truly a noble object," and he noted that it appeared as, "two individuals, both a high ruddy or orange color, though that of the smaller is of a somewhat more somber and brownish cast."

By the time Richaud made his discovery of the double nature of α Centauri, the telescope had been in use for about 80 years, and it is a little surprising that this discovery was so long in the making. French astronomer Jean Richer records making telescopic observations of the sky, and specifically α Centauri, from Cayenne in Guiana in 1673, and Edmund Halley records observing the star from the island of St. Helena in 1677. Neither Richer nor Halley, however, mentioned that it could be resolved into a binary. The ability to resolve detail in a telescope relates directly to the diameter D of its objective – mostly lenses in the case of the early observations as opposed to the more common mirrors in the present age.

At visible wavelengths of light the theoretical resolution of a telescope is expressed by the relationship $R(\text{mas}) = 1.386 \times 10^{-4}/D(\text{m})$, where the resolution is expressed in units of milli-arc seconds and the objective diameter is given in meters. This formula relates to what is called the Airy disk, named after astronomer George Biddell Airy, who first derived its properties, and the idea is that if two point sources of light are closer together than the resolution R, so they will appear as a single point source to the observer's eye.

Not much is known about the telescope that Richaud was using while at Pondicherry, other than it had a 12-ft focal length[10] and that it was apparently acquired in Siam (modern-day Thailand) in 1688. Richaud wasn't able to physically measure the separation of α Cen B from α Cen A with his telescope, but present-day calculations would make them about 7 arc sec apart, so a telescope with an objective diameter larger than a few centimeters should (theoretically) have been able to split α Centauri. Lenses with such diameters were certainly available to astronomers prior to 1689, but in practice what one sees through a telescope is greatly affected

[10] In the modern era the characteristics of a telescope are typically defined according to the diameter of its object lens or mirror. This makes sense since it is the size of the objective that determines the light-gathering power and resolution. Early refracting telescopes were usually defined in terms of the focal length, however, and in many cases these were many tens of feet in length. The primary reason for such long focal lengths was to reduce the image-degrading effects of chromatic aberration, and because it is more straightforward (although still a definite skill) to grind and polish long focal length lenses.

by the quality of the lens glass and by how much the objective image is magnified by the eyepiece lens. It is also affected by the brightness of the two stars, and α Centauri being close and bright results in significant glare. Indeed, to counteract this latter effect astronomer William Doberck, working from his observatory in Hong Kong, wrote in 1896 that measurements of the system, "ought never to be made at night. They should be measured in daylight, and as the definition is worst at sunrise, they should be observed in the afternoon."[11]

Although credit goes to Richaud for the first recorded observation of α Centauri being a double star, French astronomer Louis Feuillée independently recognized its duplicity while making observations from Conception in Chili in July of 1709. Feuillée was using an 18-ft focal length telescope and recorded that the smaller (that is, less luminous) star, which would in fact be α Cen B, was located more westerly in the sky than α Cen A, and that the two were separated by a distance comparable to the apparent diameter of α Cen B.

The first astronomer to physically measure the separation between α Cen B and α Cen A appears to have been Abbé Nicolas Louis de La Caille. Working from the Cape of Good Hope in 1752, La Caille determined that the two stars were separated by 20.5 arc sec in the sky. Not quite 10 years after this, in 1761, future Astronomer Royal of England, Nevil Maskelyne, additionally observed α Centauri from the island of Saint Helena (just as second Astronomer Royal Edmund Halley had done some 84 years earlier). Principally, Maskelyne had traveled to St. Helena to observe the June 6 transit of Venus, and for this he had been supplied with a 2-ft focal length telescope fitted with a micrometer for measuring angular separation. It was with this telescope that he made additional star observations and specifically found the two stars of α Centauri to be separated by 15.6 arc sec. In just 10 years the separation of the stars had measurably changed in the sky. The next step, of course, was to determine the full range of motion of the two stars, but this was a process requiring a great deal of patience, and it was not

[11] William Doberck; the quotation is taken from the paper: "On the orbit of η Coronae Borealis," *Astronomishe Nachricten*, 141, 153 (1896).

until the mid-nineteenth century before the first orbit determination of α Cen B about α Cen A was to be published.

1.9 Jewels in the Round

Simply seeing two stars close together on the sky does not immediately tell the observer anything. They need not be physically close to each other or even a gravitational pairing – they could just be a chance alignment. Time, however, will reveal all, and if two stars are indeed in orbit about a common center then eventually some sense of their relative motion will be revealed to the eye.

The idea that two stars might actually form a physical, gravitationally bound pairing was first discussed towards the close of the eighteenth century. British geologist and natural philosopher John Mitchell first discussed the idea of mutually bound stars in 1767, arguing that on statistical grounds the close pairings of many stars were more than would be expected by random chance alone.[12] Shortly thereafter, in 1779, William Herschel began observing double stars, specifically measuring the component star separations and position angles on the sky. Some 23 years after starting his observational program Herschel published in 1802 a catalog of his deductions in the *Philosophical Transactions of the Royal Society*. Here Herschel discussed his ideas on "the union of two stars, that are formed together into one system, by the law of attraction [gravity]." Although Herschel's observations clearly showed relative motion between close stars, the first orbit to be fully determined was that for ξ Ursae Majoris (located in the Big Bear's southernmost forepaw) by French astronomer Félix Savary in 1827. This system, which is known by the Latinized version of its Arabic name *Alula Australis*, is located about 30 light years

[12] John Mitchell (1724–1793) is perhaps better known in modern times for his suggestion that some stars might be so massive that their associated escape speed would be greater than the speed of light – making them what we would now call black holes. It was also Mitchell who developed the experiment, eventually carried out by Henry Cavendish, to determine the value of the universal gravitational constant ($G \approx 6.67 \times 10^{-11}$ Nm^2/kg^2).

from the Sun, has an orbital period of just under 60 years and is composed, like α Centauri, of two Sun-like stars.

In order to determine the orbit of a visual binary, observations of the position angle and the separation of the two stars need to be collected over time. The position angle is measured around the sky from the primary or brighter star to the secondary or fainter star and is expressed as the angle measured eastward from the imaginary great circle connecting the primary star to the north celestial pole. The separation is further measured as the linear angular distance of the two stars in the sky. As time goes by the two angles will vary, tracing out an ellipse upon the sky, with the orbital period of the two stars being revealed once the position angle begins to repeat (see Appendix 3). The first separation and position angle measurement of α Cen B about α Cen A was made by La Caille in 1752; the first observation to be made from a dedicated Southern Hemisphere observatory was that by the Reverend Fearon Fallows in 1827.

Indeed, Fallows, using British Admiralty funding, organized the construction and was the first director of the observatory at the Cape of Good Hope in Africa, established in 1821. Sadly, Fallows along with the entire observatory staff, died of scarlet fever in 1830. Also, sadly to say, the component separation deduced by Fallows was entirely spurious.

The first estimate of the orbital parameters for α Centauri B appeared in the January 12, 1855, edition of the *Monthly Notices of the Royal Astronomical Society*. Indeed, in that issue two short, half-page articles appeared back to back, with the first orbit being computed by Eyre Powell, who was working from the Madras Observatory in India. The second orbital determination was presented by John Russell Hind, who at that time was Superintendent of the *Nautical Almanac* in London, England. Hind, perhaps surprisingly, comments in his note that he had recently deduced the orbit, "not being aware that anyone else was engaged upon the same investigation." There was apparently no great haste, or even excitement, concerning the first computation of the orbit elements to α Centauri. The two orbit determinations were certainly preliminary, and were based primarily upon the observations recorded by John Herschel and Captain William Stephen Jacob (late of the Bengal Engineers).[13] Hind found an orbital period of 80.94 years and a periastron date of 1859.42. Powell found slightly different values, deducing an orbit period of 75.3 years and a periastron date of 1858.012.

The variation in these fundamental results, almost 6 years in orbital period and nearly a year and a half in the periastron date, are certainly understandable given the limited data that Hind and Powell had to work with, and the only way to improve the situation was to collect more precise positional data. Indeed, the orbit determinations presented by Hind and Powell were based upon only 21 years' worth of observations – covering a time span of about a quarter of the system's orbital period. Over the next half-century time interval more observational data were collected, with the first definitive orbit determination generally being credited to William Finsen. Working from the Union Observatory in South Africa – the same observatory from which Robert Innes discovered Proxima in 1915 – Finsen summarized his results in February 1926. The orbital period was now determined to be 80.089 years and the date of periastron 1875.7588. The position angle and radial separation diagram produced by Finsen is shown in Fig. 1.8. The ellipse that α Cen B traces out around α Cen A in the sky has a major axis (that is, its greatest diameter) of 35.33 arc sec in length. With the trajectory thus determined it appears that the angular separation of α Cen A and α Cen B in the sky (as seen from Earth) can be as small as 4 arc sec and as large as 22 arc sec.

The canonical orbit for α Cen AB in modern times is generally taken to be that presented by Dimitri Pourbaix (Université Libre de Bruxelles – see Appendix 3) and co-workers in 2002. Their data analysis reveals an orbital period of 79.91 years and a periastron date of 1875.66. The next apastron, when the two stars are at their

---

[13]Captain William Stephen Jacobs first traveled to India in 1831 and was soon subscripted into the great India survey, then under the directorship of George Everest. Retiring from the survey due to ill health, however, he turned to astronomy, eventually taking over directorship of the Madras Observatory operated by the East India Company. Jacob specialized in observations of double stars and in the computation of binary star orbits. In 1855 he suggested that the motion of the binary star 70 Ophiuchi indicated it might have a planetary companion. It was in the following year, in a letter written to the editor of the *Monthly Notices of the Royal Astronomical Society* in London, that he made the same claim for α Centauri. In each case, however, it turned out that there were no specific planets – at least as envisioned by Jacob.

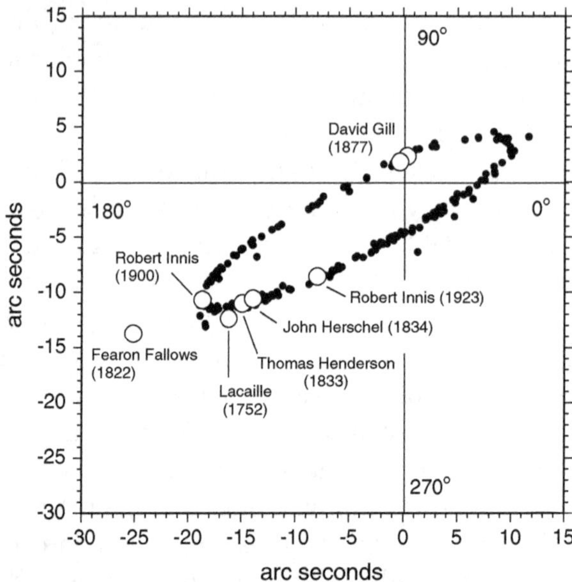

FIG. 1.8 Position angle and radial separation of α Cen A (located at the central point of the axes) and α Cen B. Each small dot corresponds to the observed location of α Cen B as recorded over the time interval from 1752 (La Caille's observation) to the end of January 1926. North is towards the right of the diagram, and the axis scale is in arc seconds (The data is taken from Circular No. 68, 1926, February 26, of the Union Observatory)

greatest physical separation, will occur in 2075. The eccentricity of the orbital path swept out by α Cen B is additionally refined to be 0.5179, and the larger diameter (major axis) of the ellipse is 35.14 arc sec across. The trajectory of α Cen B is additionally inclined and twisted in our line of sight, so that what we actually see is the apparent trajectory of α Cen B about α Cen A rather than its true orbital trajectory (for details, see Appendix 3).

The position angle and radial separation for α Cen B about α Cen A during the next century is shown in Fig. 1.9. As this book is being prepared α Cen B is moving towards periastron, its closest physical approach to α Cen A, which will be achieved on 2035.48 (June 24, 2035). The two stars will be at their closest in the sky, separated by just 2 arc sec, in 2037, and at their greatest angular separation in 2060.

There is a general human tendency to think of stars and the binary orbits of stars as occupying vast distances of space, and in general this is true. In the case of α Cen AB, however, we are dealing

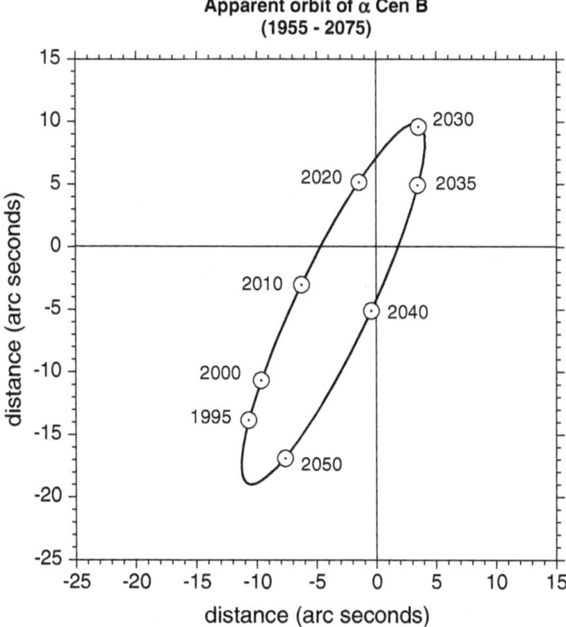

FIG. 1.9 The orbit of α Cen B about α Cen A as projected onto the sky. The next periastron passage will take place in 2035, although the closest observed approach will not occur until 2038. At this time the stars will be just 2 arc sec apart. North is towards the top of the diagram and east towards the right. The formulae used to construct the orbit and determine the separations are described in Appendix 3 of this book

with a relatively compact binary system, and it would in principal (if gravitational interactions were miraculously turned off) fit comfortably inside the planetary realm of our Solar System. Figure 1.10 shows the orbit of α Cen B as it would be traced out if α Cen A was placed congruent with the Sun. At its closest approach α Cen B would then sit just outside of the orbit of Saturn, while at its greatest retreat, it would be placed just beyond the orbit of Neptune.

1.10 The Measure of the Stars

With our description of relative separations we have moved well ahead of the history timeline relating to α Centauri. Indeed, fundamental to the development of astronomy in the first half of the nineteenth century was the development of methodologies for measuring stellar parallax.

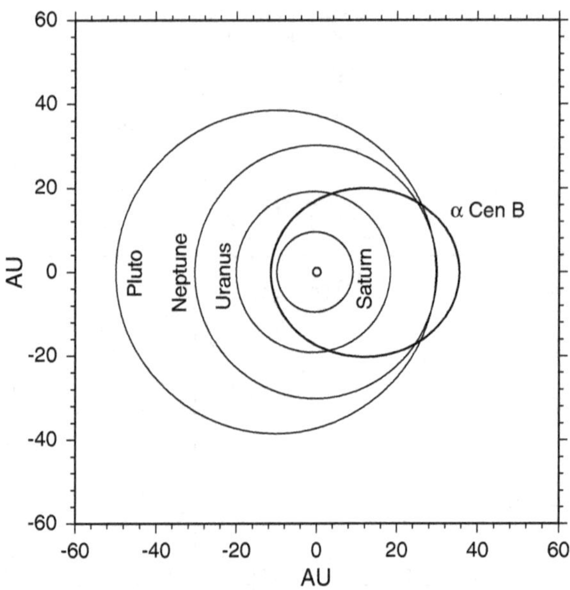

FIG. 1.10 A superposition of the orbit of α Cen B and the planetary realm of our Solar System. The *circle* at the very center of the diagram represents Earth's orbit

The idea of using a parallax shift to determine distance is many thousands of years old, and we, in fact, use it every day of our lives. It is the reason why we have two eyes set apart from each other. By measuring from two locations the slight variation in the apparent position of a nearby object, with respect to more distant ones, an estimate of distance can be made. Objects that are very close show a large parallax shift, while more distant objects show a small parallax shift.

The geometrical idea behind the measure of stellar parallax is shown in Fig. 1.11. Trigonometry provides the required relationship between the distance D to a star and the angle of parallax P, which is actually defined as half of the apparent shift in the sky in a 6-month time interval. Accordingly, $\tan P = 1\ \mathrm{au}/D$. The units are astronomical units (au) since it is the radius of Earth's orbit around the Sun that defines the baseline for the angular measurements. Like a giant conical pendulum, Earth orbits the Sun, and parallax measurements can be envisioned as finding the distance from the Sun to the pendulum's apex point centered on a nearby star.

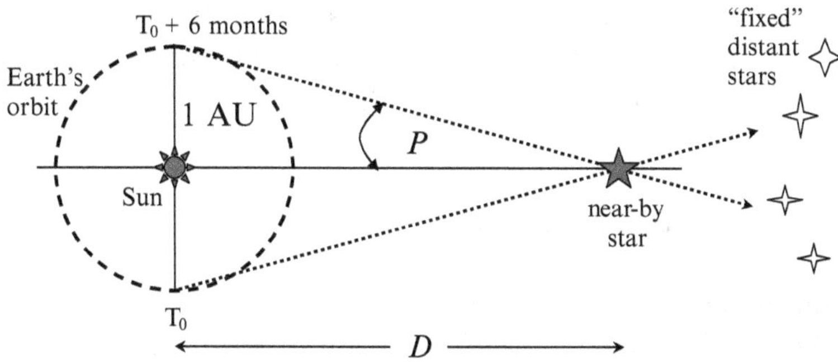

FIG. 1.11 The relationship between the angle of parallax (P), produced by the motion of Earth over a 6-month time interval, and the distance (D) to a nearby star. The more distant 'fixed stars' only appear 'fixed' because they are too far away to produce a measurable parallax (formally, as $D \to \infty$ so $P \to 0$)

By the mid-point of the eighteenth century, the Reverend James Bradley had pushed the available technology to its very limits. Writing to Edmund Halley in 1727 Bradley indicated that he was of the opinion that he could measure star displacements down to a limit of one second of arc. And, while Bradley clearly demonstrated the existence of Earth's motion around the Sun through the discovery of stellar aberration, he at no time claimed to have had measured stellar parallax. In other words, with respect to the stars that Bradley had studied, none were closer than 206,265 AU – the distance at which the angle of parallax would be 1 s of arc (see below). Even the closest of stars to our Sun, on this scale, must be an extremely remote object.

Although Bradley found no stars with a parallax larger than 1 arc sec, his limit was later used to define the standard unit of star distance measure – the parsec. The word parsec itself betrays this connection and is a concatenation of the phrase 'an angle of **par**allax corresponding to one **sec**ond of arc.' This convenient measure of distance was proposed by Herbert Hall Turner, Savilian Professor of Astronomy at Oxford University, in 1913. However, the expression and its definition were not an instant hit, and it took nearly two decades before the parsec was officially recognized, by the assembled astronomers at the first meeting of the International Astronomical Union in Rome in 1922, as the preferred unit of astronomical measure. With the parsec so defined, the distance to a star is easily computed in terms of its measured parallax via the relationship: D (pc) = $1/P$ (arc sec).[14]

1.11 Parallax Found

The historical problem with the determination of stellar parallax has been the sheer difficulty of measuring extremely small angular shifts over an extended period of time. Although Bradley found no evidence for a shift in the star γ-Draconis (also known by its Arabic name of *Eltanin*) greater than one arc second, other astronomers, throughout the eighteenth century, were less cautious and claimed to have measured sensible parallax shifts to various stars. They were all wrong, however, and their claims were generally dismissed

[14]With reference to Fig. 1.11 we have already written down that the distance to a star with parallax P is tan P = 1 au/D. Simple algebra now gives D = 1 au/tan P, and according to the parsec definition we find D = 1 pc ≡ 1 au/tan 1″.0. In the modern computer-dependent world this is where the calculation would stop; the distance would be determined by simply inputting the number for a measured parallax into a calculator. Going back just 50-years from the present, however, electronic calculators were rare things; indeed, there were only a handful of computers in the entire world. Historically, say, going back 100 years, the evaluation of a tangent and then its inverse required the use of mathematical tables, a pen, some paper and a lot of hard graft. The calculation can certainly be made, but it is tedious and time consuming. Astronomers, like most people, being conscious of both time and effort, wondered, therefore, if there might be some shortcuts to the mathematics. Luckily there is a shortcut, but before taking this shortcut, we must first consider some new mathematics, since the useful dodge that can be employed relates to angles measured in radians. The radian measure is simply another way of measuring angles, and by definition there are 2π radians in a circle. The connection with circles is revealed, of course, by the number π appearing in the definition, and the 2π term comes about since it is the circumference of a circle of radius $r = 1$. The important and useful point about radians is that when the angle φ, measured in radians, is very small, then the tangent operation simplifies to $\tan \varphi \approx \varphi$ (radians), with the approximation becoming better and better as φ gets smaller and smaller. The point of all this is that mathematically speaking, for small angles we can now write D = 1 au/P (radians), and this saves us from having to calculate the tangent of an angle. (Remember this was useful when there was no such thing as an electronic calculator.) The practical problem for astronomers, however, is that it is not possible to construct a measuring scale in radians. This, however, is not an insurmountable problem in that the angle of parallax can be measured in arc seconds and then mathematically converted to radians. In this manner, given that there are 2π radians in 360° and 3,600 s of arc in 1°, so 1 s of arc = $2\pi/(360 \times 3{,}600) \approx 1/(206, 265)$ radians. We now recover the result D = (206,265) × 1 au/P (arc sec) ≡ 1 pc/P (arc sec). From the latter relationship it can be seen that when the angle of parallax is 1 s of arc, then the distance is, by definition, 1 parsec (1 pc), and that 1 pc is equivalent to a distance of 206,265 AU.

by other observers as being unproven. This all changed, however, in 1838 when, within a just few months of each other, three astronomers, all working independently, published convincing research papers indicating that the stellar parallax barrier had finally been breached. The true distances to three nearby stars (the Sun being excluded here) were now determined, and the isolation of the Solar System was indeed recognized.

Working on the faint, but just visible to the unaided eye, star 61 Cygni Friedrich Bessel was the first to go into print. He chose 61 Cygni because it was known to have a large proper motion (to be described shortly), and this he correctly reasoned indicated it must be relatively close to the Sun. After a painstaking series of observations and equipment refinements, Bessel found a parallax of 0.2854 arc sec for 61 Cygni, indicating that it was 3.5 pc distance from us (a distance of about 723,000 au).

Next to publish was Scottish astronomer Thomas Henderson. Working at the Cape Observatory in the Southern Hemisphere, Henderson in fact studied α Centauri. Complaining bitterly of the physical conditions at the observatory, however, Henderson gathered his data and then hastily returned to Edinburgh to reduce his numbers. His observations revealed a parallax of $P = 0.7421$ arc sec, indicating that α Centauri was some 2½ times closer to the Sun than 61 Cygni, having a displacement of just 1.35 pc.

The final member of the victorious parallax triumvirate was Russian astronomer Friedrich Struve. Working from the Dorpat Observatory in Estonia, Struve studied the bright Northern Hemisphere star Vega. This star he found had a parallax of 0.129 arc sec, indicating a distance of 7.75 pc from the Sun.

Finally, in the closing days of the mid-nineteenth century, the vast openness of space was being constrained by direct geometric measure.

1.12 Thomas Henderson: The Man Who Measured α Centauri

The thought of becoming His Majesty's Astronomer at the Cape of Good Hope, South Africa, did not fill Thomas Henderson with any sense of immediate joy. Indeed, it was only the insistence of his close friends that convinced him to accept the offer of this scientifically significant but greatly underfunded post in 1831. History reveals, however, that for Henderson the move was ultimately a good one, and it resulted in his ticket to astronomical fame. Although he endured less than 13 months in the "dismal swamp" of what is now the resplendent city of Johannesburg, Henderson did managed to collect the data necessary to gauge the distance to the Sun's nearest stellar companion system

Thomas James Alan Henderson was born in Dundee, Scotland, on December 28, 1798. He was a child prodigy showing, at an early age, a great aptitude for mathematics and science. His early career, however, began with an apprenticeship to a Dundee Law firm – a position that eventually resulted in his moving to Edinburgh in 1819 to complete his studies.

The move to Edinburgh ultimately proved more significant to Henderson's astronomical interests, although his legal talents were soon spotted and he rose through the ranks to become Legal Secretary to the Lord Advocate. It was access to the observatory on Edinburgh's Calton Hill that re-ignited Henderson's interests in astronomy. It was there that he was encouraged to use the observatory's instruments, and that he began to turn his thoughts to topics in mathematical astronomy.

Traveling frequently to London on legal business, Henderson befriended some of the most prominent astronomers of the time, including George Biddell Airy (the future Astronomer Royal, 1835a, 1881), John Herschel and (more infamously) Sir James South.[15] His first scientific paper was published in the December 12, 1828, issue of the *Monthly Notices of the Royal Astronomical Society*, and it was concerned with the prediction of lunar occultations of the star Aldebaran (the brightest star in the constellation of Taurus). In 1830 Henderson further communicated to the Royal Astronomical Society a paper relating to the general prediction of occultations. He additionally published at this time a series of his observations relating to Comet Encke – a comet that he would observe on numerous future occasions.

[15] Sir James South (1785–1867) was one of the founding members of the Astronomical Society of London – later to become the Royal Astronomical Society. South was involved in a veritable soap opera-like lawsuit with famed instrument maker Edward Troughton concerning the construction of a new observatory.

Also in 1830, and importantly for the unfolding of future events, Henderson compiled a list of Moon-culminating stars for famed Arctic explorer Sir John Ross. These stars could systematically be observed by Ross to determine his location and most importantly his longitude. Henderson's astronomical skills were becoming well recognized, and though his applications for a professorship in astronomy at Edinburgh University, and as a superintendent at the *Nautical Almanac* office, were unsuccessful, he was offered the directorship of the Cape Observatory in 1831 upon the death of the indomitable and founding observatory astronomer Fearon Fallows.

Although the directorship was a means of entry into the select arena of nineteenth century professional astronomy, Henderson was reluctant to leave Scotland, and not unreasonably was worried about his chances of surviving very long in such a difficult location. Indeed, it appears that Henderson, who arrived at the cape on March 22, 1832, detested his surroundings from the beginning. This being said, he set to work on an impressive series of observations, measuring the positions of many hundreds of southern stars and the transit of Mercury on May 5 (1832). In addition he gathered observations relating to stellar occultations, parallax observations of the Moon as well as Mars, and he made observations of comets Encke and Biela. This impressive amount of work was achieved within a year and half of his arrival in South Africa. All was not well, however, and after being refused a request for funds to improve the observatory and its newly introduced time service, Henderson resigned his directorship in May 1833, returning with all due haste to his beloved home of Edinburgh.

On October 1, 1834, Henderson was elected professor of astronomy at Edinburgh University and first Astronomer Royal for Scotland. Under these titles he further took on the directorship of Calton Hill Observatory. Henderson's contributions to astronomy were recognized through his election as a Fellow of the Royal Society on April 9, 1840, but he lived a quiet life, and upon his death on November 23, 1844, he was buried in what is now a relatively forgotten corner of Greyfriars Churchyard. There is no memorial tablet to commemorate Henderson's accomplishments, and no formal portrait of Henderson is known to exist.

If it was not for his work relating to α Centauri Thomas Henderson would, by now, be an almost entirely forgotten figure – a footnote participant of nineteenth century astronomy known only to a few historians of science. As it is, however, Henderson was among the first pioneering practitioners to bag one of astronomy's biggest prizes – one for which astronomers had been searching for many centuries.

The story, as far as Henderson goes, began with the arrival of an unexpected letter, received just a few weeks before he embarked on his voyage home from South Africa in 1833. The surprise missive that set Henderson on his path towards fame was written by Manuel Johnson, a member of the St. Helena Artillery (a garrison established to protect the gravesite of Napoleon Bonaparte) and an avid astronomer. Johnson had lived on the remote South Atlantic Island for many years and had become a friend of Fearon Fallows, and had in fact visited the Cape Observatory on a number of occasions.

Importantly, Johnson had been making careful observations of the southern stars and had noticed that upon comparing his positional measurements for α Centauri with those obtained by Abbe Nicolas Louis de La Caille some 80 years earlier, in 1751, found that it must have a large proper motion (amounting to some 3.6 arc sec per year). This high proper motion, Johnson pointed out in his letter to Henderson, suggested that α Centauri was most likely a nearby star for which a parallax displacement might just be measurable. Since Henderson had gathered a series of about 100 new observations on α Centauri, spaced over a 1-year time interval while working at the Cape, he decided to use the time on his voyage home to reduce the relevant data. Remarkably, the positions of α Centauri showed a residual 'error' of about one arc second. This, however, was no real error of the observations, but the very first detection of an actual parallax. The distance to α Centauri had been gauged, and it was located about 3.25 light years away from the Sun.

With Henderson's return to England in late 1833 the story falters. At this stage in the saga, Henderson had the first accurate and not unreasonable dataset to indicate that a true stellar parallax had been determined, and yet he hid the result away for the next 6 years. The news that the parallax of α Centauri had actually been measured was first made in the January 1939 issue of the *Monthly Notices of the Royal Astronomical Society*. Almost apologetically, Henderson comments that the measurements had not been made for the purpose of attaining a parallax, and that had he been aware of the star's large proper motion earlier he would have made a greater number of observations. The article concludes that, "If we suppose the two stars are at the same distance, then the parallax = + 1″.16, with a problem error of 0″.11. It therefore appears probable that these stars have a sensible parallax of about one second of space."

Why did Henderson delay in announcing his results? The answer, it turns out, is a little perplexing. Certainly Henderson had reason to be cautious, and technically he only had the declination part of the parallax measurements reduced by 1833. The declination data had been obtained directly by Henderson, while the corresponding right ascension measurements had been made by his observatory assistant Lieutenant William Meadows. This being said, Henderson did have the entire dataset with him in 1833, but it seems that rather than reduce it all, to corroborate his declination result, he set about reducing the entire Cape Observatory dataset instead. This action rather suggests that Henderson had little faith in his parallax result. What ultimately prompted Henderson to finish and publish the data reductions for α Centauri was the announcement, in 1838, by Friedrich Bessel that a parallax for the binary star 61 Cygni had been successfully determined.

With the initial announcement made, Henderson sent a letter to his friend and successor as Cape director, Thomas Maclear, urging him to obtain as much additional data on α Centauri as he could. In due course Maclear provided a new set of positional measurements, and Henderson set about the data reduction. Writing in the April 1842 issue of the *Memoirs of the Royal Astronomical Society* Henderson presented the new results and provided a revised parallax of 0″.9128 with an error of just 0″.0640. This new parallax put α Centauri just a little further away from the Sun, at a distance of about 3.573 light years – a value, in fact, remarkably close to the modern-day distance estimate of 4.366 light years.

The total sky motion of α Cen A and B is now revealed as being decidedly complicated. Indeed their motion is a veritable celestial waltz and promenade. In addition to the continuous

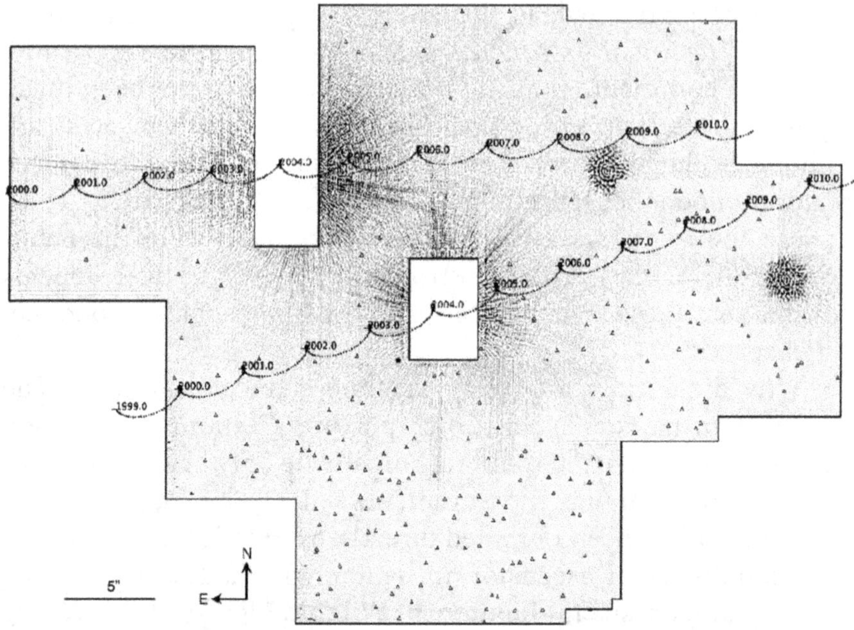

FIG. 1.12 The proper motion tracks, from 1999 to 2010, of α Cen A and α Cen B across the sky. The loops between the year positions are the result of a parallax shift, while the tracks converge towards the upper right because of the binary companionship and changing orbital separation of α Cen A from α Cen B (Image by Kervella et al. "Deep imaging survey of the environment of α Centauri. I. Adaptive optics imaging of a Cen B with VLT-NACO" (Astronomy and Astrophysics, **459**, 669 (2006)) and courtesy of *Astronomy and Astrophysics* journal. Used with permission

proper motion of the system across the sky there is now the 6-month parallax shift due to Earth's motion about the Sun to consider as well as the longer term, 79.1-year, relative shift of the two stars due to their binary motion. The combined sky motions of α Cen A and α Cen B across the sky, from 1999 to 2010, has been beautifully re-created by Pierre Kervella (Observatoire de Paris) and co-workers[16] and is shown here in Fig. 1.12.

In addition to observing and providing additional observational data on α Centauri, Maclear had earlier provided Henderson with

[16] See also, Kervella, P., and Thévenin, F. "Deep imaging survey of the environment of a Centauri. II. CCD imaging with the NTT-SUS12 camera" (*Astronomy and Astrophysics*, **464**, 373, 2007).

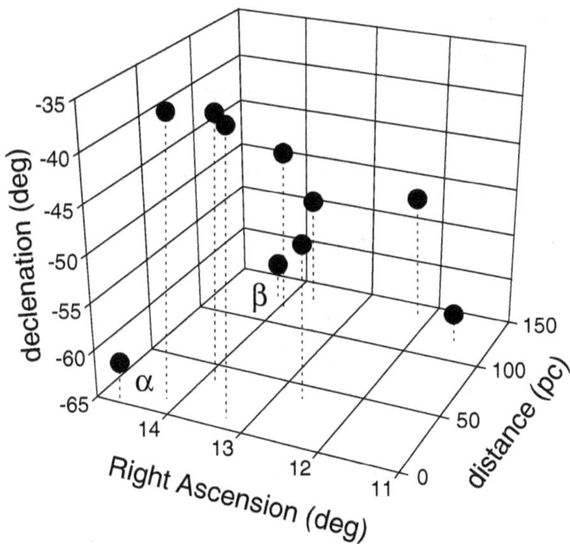

FIG. 1.13 The three-dimensional structure of the Centaurus constellation

positional data on the bright star Sirius. Indeed, not quite 1 year after his initial announcement on the parallax for α Centauri Henderson deduced a parallax for Sirius. Writing in the December 14, 1839, issue of the *Monthly Notices,* Henderson writes that, "It may be concluded that the parallax of Sirius is not greater than half a second of space, and that it is probably much less." The formal value given by Henderson was a parallax of 0.23 arc sec, which is impressively close to the modern-day value of 0.379 arc sec. Again, in December of 1842, Henderson reported upon parallax reductions relating to the southern star data sent to him by Maclear. Among the 20 stars Henderson considered in his paper is the second brightest star in the constellation of Centaurus (=β Centauri = Agena), and for this star Henderson found a parallax of 0.28 arc sec. This parallax places β Centauri at a distance some three times further away from the Sun than α Centauri. With this result, however, the limits of the then available technology had been exceeded, and the modern-day parallax for Agena is very much smaller than that found by Henderson. The parallax is actually 0.009 arc sec, and β Centauri is nearly 100 times further away from the Sun than α Centauri, making it, in fact, the intrinsically brighter object.

The three-dimensional structure of the Centaurus constellation is revealed in Fig. 1.13. The range of distance to the principle stars varies from 1.347 pc for α Centauri to 131.062 pc for ε Centauri. Remarkably it appears that the bright pairing of The Pointers, α and β Centauri, reveals an inverse intrinsic brightness; β Centauri is actually more luminous than α Centauri by a factor of about 27,400, but the additional factor of 100 in its distance dictates that α Centauri wins out on the perceived magnitude scale. Agena hides its greater light within the vastness of its distance from us.

After α Centauri the next closest set of Centaurian stars to the Sun are θ and ι Centauri, which are located just over 18 pc away; the remaining principle stars of Centaurus fall between distances of 40 and 128 pc. There is, of course, nothing unique or even unusual in this spatial separation of stars within Centaurus. Constellations, after all, are just the perceived groupings of stars on the celestial sphere. This being said, the more distant members of Centaurus are part of the so-called Scorpius-Centaurus association.

The Sco-Cen association is the nearest, and one of the most recently formed, collection of stars to the Sun. Indeed, stars are still forming within the Sco-Cen association. In total the association contains many thousands of stars, ranging in mass from just a few tenths to fifteen times (and possibly more) than that of the Sun, all moving with a common proper motion of about 0.03 arc sec per year.

Most of the stars in the body and head of Centaurus are located in the so-called Upper Centaurus-Lupus subgroup, and these stars are thought to have formed as recently as 15 million years ago. Indeed, when these stars first appeared the dinosaurs on Earth had been extinct for 50 million years. Agena, β Centauri, the second star in The Pointers, is now further contrasted against α Centauri. Not only is it much further away, and substantially more luminous than α Cen, it is also at least six billion years younger.[17] Just as in everyday life, things are not always as straightforward as they might at first appear.

---

[17] The various age estimates for α Centauri will be discussed in greater detail later.

1.13 The Discovery of Proxima

It is a remarkable fact that the closest star to the Solar System, at the present epoch, is not visible to the naked eye. Indeed, with an apparent magnitude of +11.05 a telescope having an objective diameter of at least 5 cm is required to see it under ideal conditions, and even then, picking Proxima out from the myriad background stars, would still be a challenge. Proxima Centauri is far from being a distinctive object. Faint and undersized, there is nothing about the star that immediately distinguishes it as being special. The extraordinary closeness of Proxima was revealed only slowly and via hard systematic work. Indeed, the special status of Proxima Centauri was first revealed through its proper motion, and the person to tease out this secret was Scottish-born astronomer Robert Thorburn Innes.

Working from the Union Observatory in South Africa, Innes announced the discovery of "A Faint Star of Large Proper Motion" in Circular No. 30, October 12, 1915. The discovery details take up less than one page of the publication and are based on a close examination of two photographic plates, one taken on April 10, 1910, and the other on July 30, 1915. The plates, literally large glass slides, had been obtained by staff astronomer H. E. Wood, and they showed the same star field in the vicinity of α Centauri. The reason that Innes chose these two specific plates was to look for proper motion "jumps." That is, by carefully lining up the star images from each plate in a device called a blink comparator, it is possible to detect any stars that have moved in the time interval between which the plates were exposed. By alternately switching the image presented to the operator's eye, any star that has moved between the two plate exposure times literally appears to jump backward and forward according to which plate is being viewed (see Fig. 1.14). Stars with a large proper motion will appear to jump further than those with a small proper motion, but by dividing the jump distance, expressed in arc seconds, by the time interval between the plate exposures, the sky motion in arc seconds per year can be determined. Using a blink comparator takes patience. The process is exacting, slow and tedious. Innes, however, had set himself the task of searching a region of about 60 square degrees

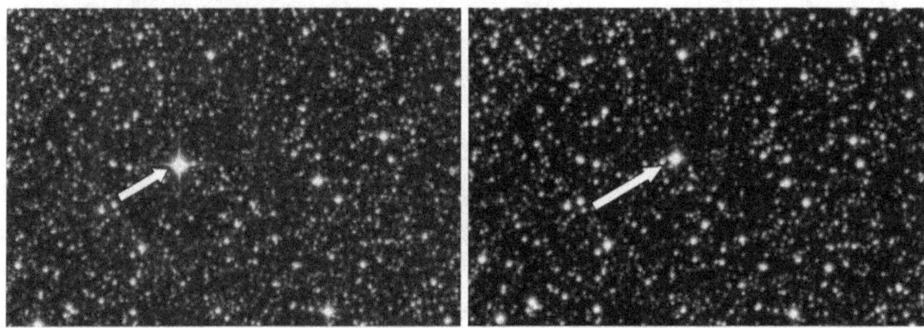

FIG. 1.14 The proper motion of Proxima Centauri (*arrowed*) is distinctly visible in these two images taken 1 year apart (Image courtesy of European Southern Observatory and the Digital Sky Survey)

on the sky around α Centauri, and the task took him about 40 h. Noting that "the strain on the eye is pretty severe," Innes explained that his many hours of plate surveying were actually spread out over a 2-week time interval. The long labor was not in vain, however, and a new, faint and previously unrecorded star was found to show a telltale proper motion jump.

For his newly discovered faint star, Innes derived a proper motion value of 4.87 arc sec per year towards a position angle of 289°.2. With this result in place Innes immediately noted, "It will be recalled that these determinations are not greatly different from the proper motion of α Centaurus itself." In other words, what Innes was drawing attention to was the fact that even though the new faint star was slightly over 2° from α Centauri in the sky, it was apparently moving through space at the same speed and in the same direction.

With the proper motion of Proxima established, the next step was to determine its parallax, but Innes and the Union Observatory did not, at that time, have the required equipment to make such measurements. With the required apparatus on order, however, Innes was presumably dismayed to hear some 18 months after the publication of his discovery article that a parallax for the star had been determined by Dutch astronomer Joan Voûte.[18]

[18]Joan George Eradus Gijsbertus Voûte is a rather obscure figure. It is known that he was an assistant at the Cape Town Observatory, and that he later took charge of the time-keeping section of the Meteorological Office in Jakarta, Indonesia. A brief obituary is provided in the *Quarterly Journal of the Royal Astronomical Society* (**5**, 296–297, 1964).

Working from the Royal Observatory in Cape Town Voûte published his results in the *Monthly Notices of the Royal Astronomical Society* on June 7, 1917. Through the analysis of multiple photographic plates Voûte had determined a proper motion of 3.67 arc sec per year, a position angle of 282°.7 and a parallax of 0.755 arc sec. With the parallax values as they were then known, Voûte's result placed Proxima 0.007 pc (1,444 AU) further away from the Sun than α Centauri. Voûte also noted that the difference between the photographic and visual magnitudes relating to Innes's faint star indicated that it must be distinctively reddish in color and was most likely, therefore, a low temperature star of spectral type M (for details see Appendix 1). With Voûte's paper in place there was little doubt that there was something highly interesting about Innes's new star. Not only was it moving at the same apparent speed across the sky as α Centauri, it was moving in the same direction and was at the same distance away from the Sun. It had to be a companion to its brighter stellar cousin. The question that Voûte then asked is one that astronomers are still struggling to determine to this very day – are they physically connected or members of the same drift? By this Voûte was asking if Innes's star was a gravitationally bound trinary companion to α Centauri, or was it simply a fainter star, possibly formed from the same birth cloud as α Centauri moving through space in the same direction. The details of this question will be further explored in a later section, and for now we leave it as an unanswered question.

Not to be outdone and presumably with a desire to stake his discoverer's claim, Innes published his own parallax for Proxima in *Circular No. 40* of the Union Observatory. This study was based upon observations gathered between May 26, 1816, and August 23, 1917, and the reductions indicated a parallax of 0.82 arc sec. With this result established, Innes noted, "It would, therefore, appear that the faint star is actually the nearest known star to the Solar System." Given the various uncertainties in the measurements available at that time, Innes was essentially guessing that his new faint star was closer to the Sun than α Centauri. His data indicated it was closer by 0.088 pc (18,151 au). Taking his parallax result at face value, however, Innes concluded his article with comment, "If this small star had a name it would be convenient – it is therefore suggested that it should be referred to as Proxima Centaurus."

Although Centauri rather than Centaurus is now generally used in conjunction with the name Proxima, the nearest star to the Sun first received its name in print on Wednesday, September 5, 1917.

Although the pioneering studies by Voûte and Innes certainly established the close companionship between α Centauri, Proxima and the Sun, more than a decade was to pass before the first highly accurate (that is, with a small probable error) parallax values were to be made available. Using the 26-in. refractor at the Yale Southern Station in Johannesburg, Harold Alden deduced parallax values of 0.762 ± 0.009 arc sec for α Cen A, 0.747 ± 0.009 arc sec for α Cen B and 0.783 ± 0.005 arc sec, for Proxima. These results confirmed Innes's guess and truly identified Proxima as being the closest star to the Sun at the present epoch. The modern Hipparchos satellite-based parallax measure for Proxima is 0.7723 seconds of arc, indicating a distance of 1.295 pc, making it just 0.005 pc (about 11,000 au) closer to the Sun than α Centauri.

1.14 The World in a Grain of Sand

Can the sheer vastness of the parallax numbers, as discovered for α Centauri and Proxima, be realized in the mind's eye? Well, of course, the answer to this question is both yes and no, but to underscore the famous lines, so aptly penned by the late Douglas Adams, "Space is big," let us imagine the following experiment, and literally describe the scale of interest in terms of a grain of sand.

Imagine that the Sun is a small grain of sand (say about 1 mm across). On this scale Earth is about 0.01 mm across – that is, 109 times smaller than the Sun grain and about 50 times smaller than the full stop at the end of this sentence. Earth's orbit around the Sun, in contrast, would be a circle with a diameter of 21.5 cm, which is about the diameter of a typical Frisbee. On the same scale as our Sun grain, α Centauri is located about 29 km away. Being Sun-like stars, both α Cen A and α Cen B will be about the same size as our Sun grain – just 1 mm across. The two sand grains that constitute our α Cen AB binary would, at their closest contact, be 1.2 m apart (at their greatest separation they would be 3.8 m apart). Again, on our Sun grain scale, Proxima Centauri would be situated about 1.2 km from the two α Cen AB sand grains. The Proxima Centauri sand grain would be about a tenth the size of the grain representing the Sun, making it about 0.1 mm across, or about a fifth the size of the full stop at the end of this sentence.

The stars, as our sand-grain scale tries to indicate, are widely spaced. The closest star to the Sun is 29 million Sun-diameters away, and the imagined scale is that of several small sand grains separated by a distance of 29 km (3.4 times larger than Mount Everest is high). Space is indeed big, and perhaps even more surprisingly to our mind's eye view, space is mostly devoid of stars. Remarkably, even though our Milky Way Galaxy contains of order 300 billion stars, if you were suddenly transported to some random location within its disk the odds of you actually materializing within a star are a very healthy 100,000 billion billion to 1 against – which is not to say that materializing in the near absolute zero temperature and almost perfect vacuum of the interstellar medium would be beneficial to one's extended life expectancy.

1.15 Robert Innes: The Man Who Discovered Proxima

Death often strikes in a sudden and entirely unexpected manner. And so it was, in this very way, on March 13, 1933, that Robert Thorburn Ayton Innes took his final breaths. It was the end of the man who had discovered Proxima Centauri. In the peaceful surroundings of Surbiton, England, a massive heart attack brought to a close the fruitful and distinguished life of a talented astronomer and mathematician.

Just days before his passing, in apparent rude health and characteristic jovial mood, Innes had joined in the celebrations marking the retirement of the ninth Astronomer Royal, Sir Frank Dyson. Innes was in the good company of his peers at Dyson's farewell, and although having retired himself 6 years earlier he had completed 24 years as director of the Transvaal (later Union) Observatory in South Africa. During his time as director, Innes had become well known for his extensive work on binary stars, tracking the moons of Jupiter, and determining stellar proper motions and distances.

Born in Edinburgh, Scotland, on November 10, 1861, Innes was the oldest of twelve children born to John Innes and Elizabeth Ayton, and although he had very little formal education, he demonstrated at an early age a clear mathematical talent. By age 17 years he was elected a Fellow of the Royal Astronomical Society, and judging from his first recorded correspondence in the *Astronomical Register* for May 1878 he was already interested in and fluent with the literature and theory relating to the dynamics of the Jovian moons – a topic that he would pursue, in fact, for much of his life.

Although it is to be assumed that Innes continued his mathematical readings and interest in astronomy, we do not find him publishing again until 1889. By this time Innes had not only entered into wedlock, marrying Anne Elizabeth Fennell in 1884, he had also emigrated to Australia and established himself as a wine and spirits merchant. His second publication appeared in the August pages of the *Monthly Notices of the Royal Astronomical Society*, and consisted of a mathematical note on La Verrier's Tables du Soleil. Indeed, Innes had found a small mathematical error in La Verrier's analysis – a point that John Couch Adams verified in a commentary note directly following Innes's analysis.

Not only did Innes continue his mathematical studies in Australia, he also began to develop an interest in astronomical observing. In 1899 he wrote to *The Observatory* magazine describing a visit to the observatory of John Tebbutt, a renowned amateur observer and finder of new comets. Shortly after this time Innes was loaned a telescope by another highly regarded amateur observer named Walter Gale, and with this instrument he began a series of observations relating to binary stars. This latter work resulted in a number of academic publications and catalogs, and indeed, so impressed was David Gill, then His Majesties Astronomer at the Cape, that Innes was invited to join the staff at the Royal Observatory in South Africa. Packing up his family and mercantile dreams, Innes moved to Cape Town in 1896, taking on the official duties of the Secretary at the Observatory. He also continued to make his own observations and began to construct the first reference catalog of southern double stars.

In 1903, again upon the recommendations of (now) Sir David Gill, Innes was appointed director of the Transvaal Observatory in

Johannesburg. At that time the Transvaal Observatory was mostly concerned with meteorological, seismological and time observations, but Innes soon established a program to measure double stars, and the phenomena (specifically transits and occultations) of Jupiter's moons. Innes was in his element, and the telescopes at the observatory were soon being used to study comets, asteroids, variable stars and nebulae. In January of 1910, Innes was the first observer to sweep up the Great Comet (C/1910 A1) as it moved towards perihelion. Halley's Comet was also carefully observed by Innes and fellow observers in Johannesburg later that same year.

The year 1912 saw the Transvaal Observatory undergo a name change, becoming the Union Observatory, and it also saw construction begin on a dome to house a new 26.5-in. refracting telescope commissioned by the South African government. The First World War and its aftermath would delay, however, the completion and installation for over twenty years; the telescope finally seeing first light in January 1925, just 2 years before Innis retired as director of the observatory. Also in 1912 the observatory acquired a Zeiss stereo-comparator, and this precipitated the initiation of a highly successful program to measure stellar proper motions – culminating, as far as we are concerned, with the discovery of a faint star (later to be known as Proxima Centauri) in 1915.

Not only did the observatory staff begin to "blink" their own photographic plates, they also analyzed a large number of plates from the Cape Observatory and some from the Greenwich Observatory in England. During these reductions a faint, spectral type M3.5 star with a record-breaking (for the time) proper motion was discovered, and this star (rather than Proxima) has become known as Innes's Star (also cataloged as HD 304043 and LHS 40). Innis's star is currently on the HARPS (Highly Accurate Radial-velocity Planet Searcher) list for study as having a possible planetary supporting system (see Sect. 2.11 to follow). In recognition of his distinguished career and fruitful collaboration program the University of Leiden in Holland conferred an honorary doctorate upon Innes in 1922. Additional recognition of his work and long-running career has been posthumously bestowed upon Innes through the naming of asteroid 1953 NA (discovered from Johannesburg by J. A. Brewer in July 1953), as well as a 47-km diameter far-side lunar crater, in his honor.

Innis retired as director of the Union Observatory in 1927 – he was then 66 years old. In this same year Innis saw the publication of his great work, the *Southern Double Star Catalog*. There was to be no sedentary retirement for Innis, however, and he continued to work on various astronomical and mathematical projects, keeping up a good publication rate until his untimely death in 1933. Much of his later work was concerned with the derivation and computation of cometary orbits, developing methods by which the various equations could be solved for through the application of mechanical calculating machines. His final publication appeared posthumously in volume 30 of the *Memoirs of the British Astronomical Association* (1935) and concerned the orbital determination of comet 1927 f (Gale) – now 34P/Gale.

The topic of this final analysis is perhaps rather fitting, since this comet was discovered by the same Walter Gale who loaned Innes an old 6.25-in. refracting telescope (built by famed instrument maker Thomas Cooke) in 1894, a few years after his arrival in Sydney, Australia. Indeed, it was with this very instrument that Innis, then a bookish mathematician, first cut his teeth as an observational astronomer.

1.16 Past, Present and Future

Given that Proxima Centauri and its erstwhile companions in α Centauri are the closest stars to the Sun, when, we might reasonably ask, did this special spatial condition come about? Likewise, when will their reign end and some other star take over the mantle of nearest neighbor? Is it possible that some yet to be discovered, faint star is in a position to usurp Proxima's special closeness condition – a new Proxima lurking unidentified, its faint glow passing unnoticed within the vast darkness of the night?

The answer to this latter question is certainly no. There are at least two arguments to bolster our confidence in this negative answer. First of all, statistically it is unlikely. This argument follows from star counting and parallax surveys, and recognizes the fact that within the solar neighborhood there are some 0.09 stellar systems per cubic parsec of space. On this basis the typical (or average) separation between star systems should be of order

2.8 pc. With this number we are already ahead of the statistical game; the odds of Proxima being just 1.3 pc away from us are essentially 1 in 10. The odds of a star being closer than 1 pc to the Sun are about 1 in 22, which are perhaps not bad odds for a horse race (provided the touted horse actually wins) but less than encouraging odds for finding a new Proxima (during the present epoch anyway).

A second argument that supports the notion that Proxima truly is the closest star relates to the intrinsic brightness of such M dwarf stars. Proxima is certainly faint, coming in at an apparent magnitude of +11.05, but, of course, if it was any closer then it would be brighter. At 1 pc away, for example, Proxima would be half a magnitude brighter than at present, while at 0.5 pc it would take on an apparent magnitude of +9, still invisible to the unaided human eye but well within the detection range of a small telescope and even a pair of binoculars.

The Hubble Space Telescope guide star catalog provides data on some 20 million stars with apparent magnitudes between +6 (just visible to the human eye) and +15. The Sloan Digital Sky Survey (Legacy catalog) further contains data on about 230 million celestial objects as faint as magnitude +22 (in an area covering about 20 % of the entire sky). These surveys do not absolutely rule out the possibility of a missed 9th or 10th magnitude star, but they certainly make the likelihood of any such object existing, and being overlooked, extremely small.

Although the likelihood of a new Proxima being found closer than 1 pc to the Sun is (at an optimistic best) extremely remote, this is not to say that closer astronomical objects might not yet be found. These potential new neighbors would not be stars, however, but free-floating Jupiters[19] and possibly their larger cousins, the brown dwarfs. The conditions for stardom will be described in the next section, and for the moment let us just state that the lowest mass object that can become a star, essentially defined as an object that generates energy within its interior through sustained hydrogen fusion reactions, will have a mass of about 0.08 times that of the Sun. Below this mass limit (equivalent to about 84 times

[19] Such planets are thought to have grown within the gas/dust disks associated with newly forming proto-stars. The (rogue) planets are then later ejected from the host system as a result of gravitational interactions with other, larger planets.

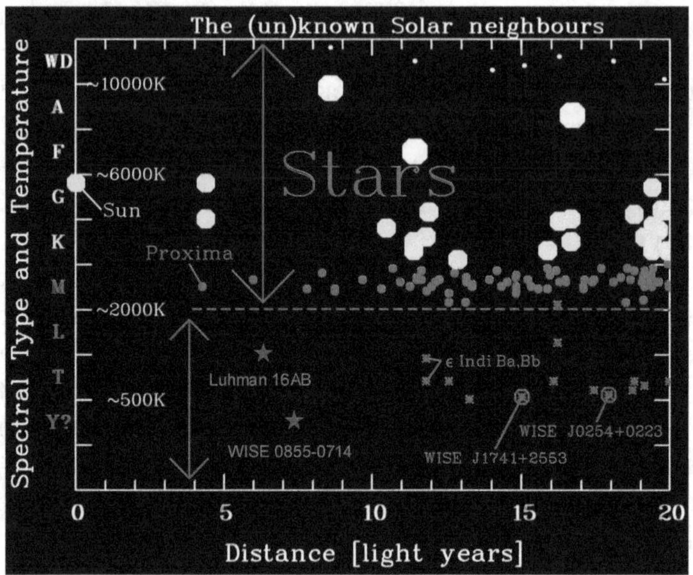

FIG. 1.15 The known solar neighborhood (Image based upon an original diagram from the Leibniz-Institut für Astrophysik Potsdam. Adapted with permission)

the mass of Jupiter) are found the brown dwarfs; so named because they have inherently low surface temperatures (smaller than a few thousand degrees) and because they are relatively small – although this being said, they have sizes similar to that of Jupiter (or about 10 times the size of Earth). Being so small, cool and of very low luminosity, brown dwarfs are extremely faint at optical wavelengths. They are, in fact, at their brightest at infrared wavelengths.

Figure 1.15 shows the distance to the currently known stars and brown dwarf objects out to a distance of 20 light years (6.13 pc) from the Sun. The closest known brown dwarfs (a binary system, in fact, with the designation WISE J104915.57-531906.1 – or more easily, Luhman 16AB) were discovered in 2012 and are located just 2 pc away, a distance less than twice that to Proxima.

More and closer brown dwarfs are likely to be found in the future, but just how close remains to be seen. Chilean astronomer Maria Teresa Ruiz discovered one of the first known brown dwarfs in 1997, and this particular object, estimated to be some 18.7 pc from the Solar System, glows at a faint apparent visual magnitude of +21.8. Bringing this same object to a distance of 0.5 pc from the

Solar System would only increase its visual brightness to about +14 – well below human eye threshold, and well below the brightness limit of virtually all large-scale stellar survey catalogs. Brown dwarfs could be numerous and very close to the Solar System and yet we will struggle to find them.

The solar neighborhood is just a miniscule component of the Milky Way Galaxy, which itself is a whirlpool of rotation and stellar motion. The Sun takes about 260 million years to complete one orbit around the galactic center, following a carousel-like path, taking it above and then below the galactic plane with a period of about 30 million years. It marches along its orbit at a speed of 220 km/s, and it is accompanied in its journey by an ever shifting cohort of mostly smaller mass, lower luminosity stars – stars, in fact, just like Proxima Centauri.

In their motion around the galactic center the stars execute a subtle quadrille, coming together and moving apart according to their specific speeds and directions of motion. Astronomers try to make sense of these maneuvers by subtracting out the motion of the Sun and referring the motion to the local standard of rest (LSR). The manner in which another star will approach, and then recede from the Sun, can then be determined according to just a few straightforward observations.

Key to knowing how close another star will pass to the Solar System is a measure of its space velocity and its motion relative to the line of sight at some specific epoch – in other words, we need to know how fast it is moving and in what specific direction. The equations that describe the observations and enable the time-dependent path of a star to be determined in the LSR are given in Appendix 2, and the reader is referred there now to take a quick look a Fig. A2. This diagram illustrates the essential geometry of the problem, provided we can determine the speed and direction of motion of a star at some specific instant so its path and location at earlier and later times can be enumerated. In this manner the time of closest approach, and the brightness of the star at that point, can be calculated. Figure 1.16 shows the approach and recession of several nearby stars over the time interval from 20,000 years ago to 80,000 years into the future.

It is not specifically clear from Fig. 1.16, but α Centauri became the closest star system to the Sun about 50,000 years ago,

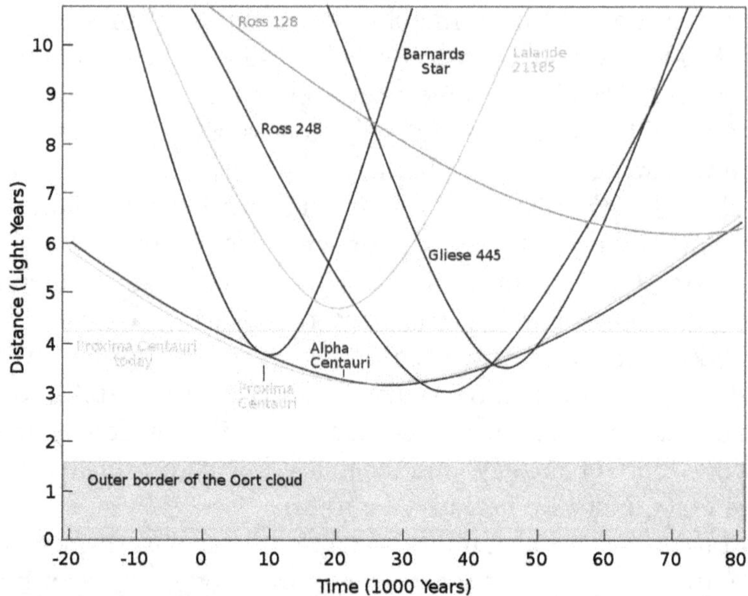

FIG. 1.16 The close approach distances of several selected solar neighborhood stars (Image courtesy of Wikimedia Commons. Near-stars-past-future-cs.svg)

and it is still moving toward its point of closest encounter – to be reached in about 23,000 years from the present. After this, the story becomes just a little complicated. In about 32,000 years from now the star Ross 248 will become the Sun's closest companion, only to be usurped about 10,000 years later by Gliese 445. During this 20,000-year time interval, however, α Centauri is still hovering close to the Solar System, and while losing its nearest neighbor crown 32,000 years hence, in about 50,000 years from now it will ascend to the throne once again and take on the status of nearest neighbor for a second term. The new reign will last for some 30,000 years, and α Centauri will only lose, once and for all, its nearest neighbor status 80,000 years from now when the star Ross 128 takes on the crown.

Reflecting on the fact that Proxima is one of the most common types of stars to be found within the galaxy, from the Sun's perspective when Ross 248, Gliese 445 and Ross 128 are its nearest neighbors little will have changed, since all three of these stars are low mass, M spectral-type red dwarf stars – just like Proxima.

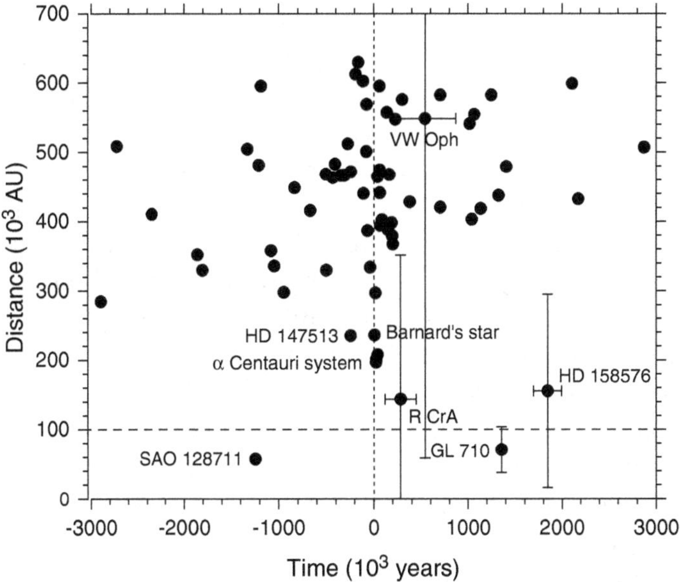

FIG. 1.17 Close stellar encounters with the Solar System from three million years ago in the past to three million years into the future. In this entire time interval Gliese 710 will undergo the closest approach to the Solar System, in about 1.5 million years time, with an estimated miss distance of 0.25 pc (62,000 au). After Gliese 710 the next closest encounter will occur about 500,000 years later – this time with the star HD 158576, a binary system in which the brightest member is a little more massive and slightly more luminous than the Sun

Technically, however, it appears that Ross 248 will approach a little closer to the Sun than α Centauri, with a miss distance of 0.93 pc as opposed to 0.95 pc. The closest approach distance of Ross 128 will be about 1.9 pc, which is two times further away than α Centauri will be from the Solar System at its closest approach.

Although no star is going to directly collide with the Sun anytime soon, it is known that a particularly close encounter, by stellar standards, is going to take place in about 1.5 million years from the present. Figure 1.17 reveals the times and closest approach distances of stars to the Solar System over the time interval three million years into the past and three million years into the future. Over the next few million years our closest shave is going to be that associated with the star Gleise 710 – another red dwarf star like Proxima. At its closest approach, estimated to be within 0.25 pc of the Sun, Gliese 710 will pass through the outer boundary of the Solar System's Oort Cloud, possibly triggering a great influx of cometary nuclei into the inner Solar System, but it will not pass close enough to gravitationally perturb the orbits of the planets.

Besides Gliese 710, two other stars might potentially perturb the Oort Cloud within the next few hundred thousand years. These are the newly formed, pre-main sequence, B spectral-type star R Corona Australis (R Cr A) and the aged red giant Mira-type variable star VW Ophiuchi. The uncertainty in the encounter conditions for these two stars is entirely related to their poorly determined parallax and proper motion characteristics at the present time, resulting in the large uncertainty bars shown in Fig. 1.17. Indeed, based upon its presently measured properties VW Ophiuchi[20] could pass as close as 0.26 pc to the Sun, or it could miss us by as much as 28.5 pc. The probability of VW Ophiuchi actually passing very close to the Solar System is fortunately decidedly low, and a straightforward analysis reveals a 1 in 8 chance of an encounter closer than 1 pc in the time interval between 300,000 and 700,000 years from the present. There is a 1 in 74 chance of an encounter closer than 0.5 pc some 350,000 years from the present time. The likelihood that VW Ophiuchi will pass closer to the Sun than Gliese 710 is in fact very low, and comes out to be something of order one in several thousand. Time, along with improved parallax and proper motion data, will ultimately determine the encounter circumstances between the Sun and VW Ophiuchi.

The star R Cr A is presently located some 8.2 pc from the Sun, and its present encounter status has recently been reviewed by Juan Jimenez-Torres (Ciudad Universitaria, Mexico) and coworkers, who derive a minimum approach distance of 0.54 pc sometime in the time interval of 100,000–500,000 years from the present. The uncertainties in the encounter conditions are truly large, but again refined future observations will eventually pin down what is going to happen. For now, we wait and watch.

[20] These results are based on a straightforward simulation of the possible VW Ophiuchi encounter conditions. The simulation considered the closest encounter distance and time of encounter for 100,000 VW Ophiuchi clones, with proper motion and initial parallax values chosen at random from within the presently allowed range of observational uncertainties.

To paraphrase the words of poet Walt Whitman, with the numbers, charts and columns now arrayed before us (specifically Fig. 1.17), it is remarkable to note that over the time interval stretching three million years into the future α Centauri will be the only system containing Sun-like stars to pass within 2 pc of us. Indeed, α Centauri is the only system containing Sun-like stars to have passed within 2 pc of the Sun during the last three million years as well. If ever humanity wanted to explore a close, second Sun-like home then surely the time is now, or at least very soon, before α Centauri leaves us hull-down, setting sail for more distant galactic seas.

1.17 Location, Location, Location

The distance versus time diagrams presented in Figs. 1.15, 1.16 and 1.17 only show the linear separation between the Sun and the present-epoch nearby stars that constitute the solar neighborhood. To complement these diagrams, the three-dimensional structure of stars located close to the Sun is shown in Fig. 1.18, and from this diagram we begin to get a feel for the real make-up of nearby space (even though it accounts for a mere 1/300,000,000 of one percent of the volume of the disk of the Milky Way Galaxy).[21]

The nearby stars are scattered more or less at random – swimming through a vast interstellar swell like so many fish upon on a galactic spiral-arm reef – and they constitute a mixed-bag of species, or spectral types. The picture of the solar neighborhood shown in Fig. 1.18, although constructed for the present epoch, should, astronomers believe, be a fairly typical description of any random location within the disc of our Milky Way Galaxy – no matter the specific time in recent galactic history one chooses. It will also remain the typical view for our Sun for many billions of years yet to come. Certainly the exact number of stars, up and down, and their locations will change, but the essential stellar

[21] This calculation assumes that the galactic disk has a ring-like structure with an inner radius of 2 kpc and an outer radius of 16 kpc, with a uniform thickness of 1 kpc. This gives a 'disk' volume of just under 800 million cubic parsecs. The solar neighborhood volume exhibited in the Fig. 1.18 amounts to that of a sphere of radius of 3.8 pc – giving a volume of just under 230 cubic parsecs.

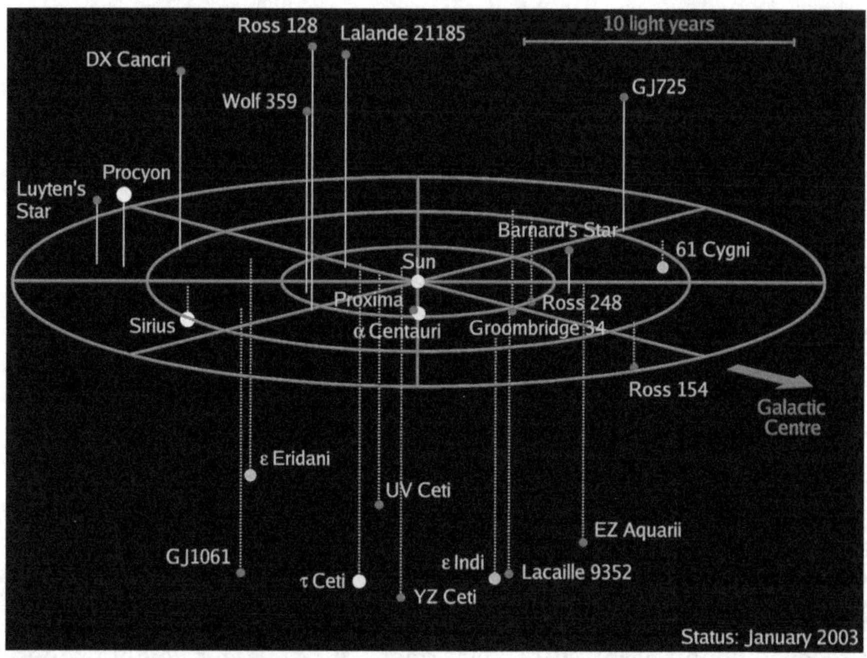

FIG. 1.18 A three-dimensional map of solar neighborhood out to a distance of 12.5 light years (3.8 pc) from the Sun. Within this volume of space there are a total of thirty-three stars (including the Sun) consisting of one A, one F, three G, five K, and twenty-one M spectral-type stars, along with two white dwarfs and at least five brown dwarfs (two of which are in orbit about ε Indi). The stars themselves include sixteen single star, four binary star and three triple star systems (including α Centauri). The number of known planets within this volume of space is fourteen (five in orbit about Tau Ceti, one in orbit about α Cen B and the eight planets located within our Solar System). The image has been labeled "status pertaining to 2003," which reflects the high potential for yet more planets and brown dwarfs being found in the future (Image courtesy of ESO)

distribution count will generally be preserved, with most of the Sun's neighbors being K and M spectral-type stars and very low luminous brown dwarfs. Extreme stars such as the O and B spectral types will only rarely come close to the Sun, since their numbers and lifetimes are miniscule when compared to their low mass, low temperature counterparts. The solar neighborhood is an essentially benign region of the galaxy and will only rarely suffer an extreme astronomical event – which is not to say that the Sun, α Centauri and the solar neighborhood haven't seen their fair share of disruptive epochs.

Having already seen that in the solar neighborhood the likelihood of very close, planetary orbit-perturbing encounters between stars, on the time scale of many billions of years, is extremely remote (see also Appendix 2 in this book), what other galactic events might be cause for concern, not only for life on Earth but for life that has chanced to evolve on any exoplanet, or associated exomoon, within the solar neighborhood? Essentially, at least for about the past five to six billion years of galactic history, the only globally destructive phenomenon of concern to planet-based life (and we know of no other) is that of a nearby supernova explosion. Even these massively energetic events, however, are apparently of no great global concern, since at least thirty supernovae must have occurred within about 10 pc of the Sun and Solar System since they first formed,[22] and yet no single catastrophic extinction event on Earth has been unequivocally linked to a supernova explosion.

This picture was likely very different in earlier epochs of the galaxy, however, when the star formation rate was much higher, and it is also dependent upon the mass and star formation efficiency of the specific interstellar cloud out of which a given star and planetary system forms. That at least one supernova must have exploded close to the solar nebula at the time that the planets were first forming is evidenced by the present-day detection of ^{60}Ni in iron meteorites. Such meteorites are derived through asteroid collisions in the region between Mars and Jupiter, and, as we shall see in Sect. 2.9 later, the asteroids are essentially remnant planetesimals – with planetesimals being the basic building-block

---

[22] The data relating to observed supernova rates indicate that something like 19 supernovae (of all types, but mostly Type II) will occur within our Milky Way Galaxy every 1,000 years. These supernovae will occur at random locations within the disc of the galaxy, and if we imagine each supernovae having a circular region of devastation with a radius of 10 pc, then something like 2.5 million supernovae would need to occur before one is likely to be placed within 10 pc of the Sun. (As with Note 21, the disk is taken to have an inner radius of 2 kpc and an outer radius of 16 kpc.) To achieve this disk coverage, and Solar System-threatening location, would require about 132 million years of supernovae explosions. Accordingly, given the Solar System is 4.5 billion years old, so of order 34 supernovae must have occurred within 10 pc of the Sun. Likewise, over the next two billion years some 15 additional supernovae will likely occur within 10 pc of the Sun. A review of observed supernova characteristics and the possible consequences of a close supernova explosion to the Solar System is given by the author in the article, "The past, present and future supernova threat to Earth's biosphere" (*Astrophysics and Space Science*, **336**, 287, 2011).

structures out of which planets are made. The ^{60}Ni found in iron meteorites (and hence within asteroids) is derived through the radioactive decay of ^{60}Fe, and this conversion process has a short half-life decay time of 1.49 million years. The only known locations where ^{60}Fe can form are within the exploding envelopes of supernova, and to explain the presence of its radioactive decay partner within iron meteorites requires that the ^{60}Fe must have been mixed into the proto-solar nebula very shortly, that is within a million years, of the supernova occurring and right at the time that the first planetesimals were beginning to form.

Astronomers distinguish between several supernovae types. Type I supernovae occur in binary systems as a result of a mass exchange between the component stars, while Type II supernovae are related to the catastrophic collapse of single massive stars. Irrespective of type and detonation physics, the essential characteristic to worry about with respect to supernovae is that of range. The closer the supernova, for example, the more severe the effects will be upon the structure of a planet's atmosphere. Unless located right on top of a planetary system, which as we have seen is an extremely unlikely event, a supernova will not cause the destruction of a planetary system. The planetary orbits will not be disrupted, nor will planets be destroyed. With respect to the persistence of life, it is the dosage of supernovae-produced gamma rays that is crucial.

A secondary effect would be the encounter of the planet with debris blast wave – the debris being the material content of the star that has undergone supernova disruption. It is estimated that Earth's biosphere will likely come to an end, and all life become extinct, in about two billion years from the present (see later). This catastrophe will be due entirely to the Sun and its ever-increasing luminosity. Nonetheless, within this same time frame something like fifteen supernovae should occur within 10 pc of the Sun. These supernovae encounter numbers will apply to all of the stars that make up the present solar neighborhood, and α Centauri, being at least 1.5 billion years older than the Sun and Solar System, should have already experienced about ten even closer (that is, within 10 pc) supernovae events.

Within about 1,000 pc of the Sun there are presently twelve pre-supernovae candidates, and of these the next most likely system to undergo supernova disruption, within the next several millions of years, is Betelgeuse (α Orion), the brightest star in the constellation of Orion. Betelgeuse, an M spectral type, Iab luminosity class supergiant star, is located about 150 pc away from us, and at this entirely safe distance, when it goes supernova, it will briefly acquire an apparent magnitude of about $m = -11$, making it a truly spectacular, but thankfully benign, cosmic display – a sparkling diamond, one might say, in the rough.

The dramatic effect that supernova explosions can have with respect to shaping the characteristics of the interstellar medium, the gas and dust between the stars, is exemplified by the solar neighborhood's current location. Indeed, the Sun, along with its retinue of stellar neighbors, are situated close to the center of the Local Bubble, a region of lower than average gas density within the interstellar medium (Fig. 1.19). This expansive region of low density dust and gas was probably cleared by a combination of supernova detonations and the strong winds produced by massive stars before they underwent supernova death. Neither the Sun nor any of the stars in the solar neighborhood were formed in the Local Bubble; they just happen to be passing through this region at the present epoch. At other times the Sun, on its journey around the galactic center, will pass through denser molecular cloud regions of the interstellar medium, as well as through spiral arm configurations.

The Sun, solar neighborhood and the Local Bubble all form part of what is called the Orion (or Local) Spur – an apparently minor spiral arm feature of the Milky Way Galaxy. Indeed, we are currently located between two much larger spiral arm features – the Perseus arm and the Sagittarius arm. The spiral arms themselves are mapped out by regions of evanescent star formation, bright galactic clusters and massive stars, and indeed, this is where the greater amount of the galaxy's luminosity resides. Although the spiral arms are continually reforming in the wake of a spiral density wave that rotates around the center of the galaxy, they are not the galaxy's dominant mass structure. Indeed, in the manner of the meek inheriting Earth, so the majority of the galaxy's disk mass is spread among the much more numerous and uniformly distributed faint stars rather than the relatively few, very luminous, more massive stars that delineate the spiral arm features.

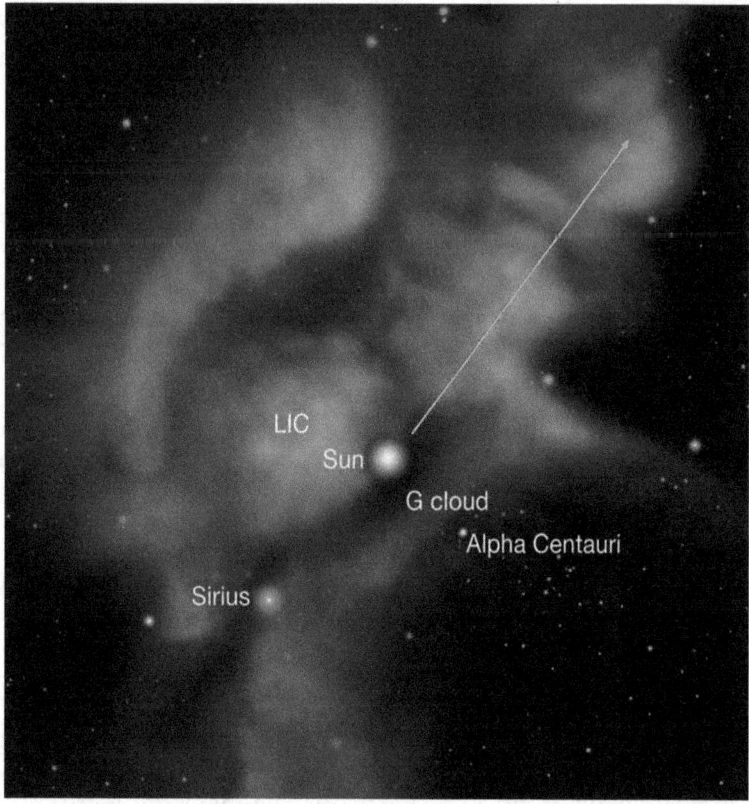

Fig. 1.19 The solar neighborhood and the surrounding interstellar medium. The Sun sits within the Local Bubble (or local interstellar cloud, LIC), which additionally stretches above the galactic plane, making in fact for a more cylindrical, chimney-like structure. Regions of low density hot gas are shown in white, while the dark areas indicate regions of higher density cooler gas. The map has sides corresponding to a scale of 50 light years (15 pc). The positions of the Sun relative to α Centauri and Sirius are indicated, and the so-called G cloud (the interstellar cloud surrounding α Centauri and Proxima) is labeled. The *arrow* indicates the direction of the Sun's galactic motion (Image courtesy of NASA)

Due to the different rates of rotation within the galactic disk the Sun typically encounters and passes through a spiral arm structure every 100 million years, and at these times the supernova encounter risk increases slightly. Some researchers have attempted to link mass extinctions on Earth, as well as terrestrial ice-age cycles, to the times of spiral arm crossing. But the arguments for these effects are far from convincing given the data that exists at the present time.[23]

Returning once more to Fig. 1.18, the stars that currently surround the Sun, next to α Centauri, are a mixture of the obscure and the famous, making for a strange menagerie of names and numbers. Indeed, many of these closest neighbors, as we have already seen, have a long and distinguished astronomical heritage; others are simply fainter lights with less important stories to tell. Among the more famous stars is the relatively bright binary system 61 Cygni, which, as we saw earlier, was the very first star to have its parallax accurately measured. Likewise, the faint M dwarf Barnard's star is distinguished as having the highest known proper motion. Discovered by American astronomer Edward Barnard in 1916, this star is currently the fourth closest to the Sun and is moving through space at some 140 km/s (over 100 times faster than a speeding bullet!), and it will be at its closest approach distance of 1.1 pc in about 10,000 years from the present. At this time both it, Proxima and α Centauri will all be at about the same distance from the Solar System, making any one of them an ideal target for space exploration. The story of the British Interplanetary Society's *Project Daedalus*, in which a space mission to Barnard's star is outlined, will be recounted later.

Luyten's star is another high proper motion star, but this time named for Dutch-American astronomer Willem Luyten. Indeed,

[23] It has been argued, for example, that during the times of spiral arm crossing not only is the supernova threat enhanced, as a result of there being more massive stars within spiral arm regions, but so, too, is the likelihood that some form of Oort Cloud disruption enhanced. This latter effect comes about due to the enhanced density of giant molecular clouds in the spiral arm. Additionally, the enhanced number of gravitational perturbations, it has been argued, should result in an enhanced cometary influx to the inner Solar System and accordingly to more impacts on Earth. Furthermore, the cosmic ray flux will be higher at times of spiral arm crossing, and this will potentially result in enhanced atmospheric ozone depletion. Another effect, first discussed by William McCrea (University of Sussex) in 1975, is that accretion of material by the Sun, if it chances to pass through a particularly dense region of the interstellar medium, could result in its luminosity increasing and precipitating thereby a dramatic warming of Earth. Although it is acknowledged that all of these various phenomena and effects could happen, there is no clear and unambiguous evidence to indicate that they actually have happened or will. Certainly the Solar System is not disconnected from events and phenomenon occurring in the rest of the galaxy, but it is far from clear what effects, if indeed any, the connections might have had upon Earth's historical past (and indeed, will have upon its future).

Luyten compiled and worked for many decades on the Bruce Proper Motion Survey catalog, starting in 1927, and he took particular interest in those stars showing the highest proper motions. These high proper motion stars are interesting since not only are they close to the Sun, but many have orbital characteristics that indicate that they must be derived from the halo region of our Milky Way Galaxy. Indeed, the oldest known star in the solar neighborhood, the Methusela star (HD 140283, presently located some 58 pc from the Sun) is a high proper motion star, with a space velocity of some 360 km/s and an estimated age of 14.46±0.31 billion years. This age estimate indicates that the Methusela star must have been one of the very first stars to have formed within the universe.[24]

Ross 248 and Ross 128 are destined, as we saw earlier (recall Fig. 1.15), to be the future closest neighbors to the Sun. These stars acquired their name and designation numbers from the catalog compiled by American astronomer Frank Ross, who worked from the Yerkes Observatory in Chicago from 1924 to the time of his death in 1939. Although not actually shown in Fig. 1.17, because it is currently 5.4 pc (17.6 light years) away from the Sun, the star Gliese 445 is additionally destined to play the temporary role of closest star to the Sun in about 45,000 years (see Fig. 1.15) from now. The Gliese (GJ) stars are so named by their placement in the *Catalog of Nearby Stars*, published by German astronomer Wilhelm Gliese in 1957.

The star Wolf 359 is named from its placement in the catalog published by another German astronomer, Maximillian Wolf, in 1919. This particular star is of interest to astrophysicists since its estimated mass makes it equal to about 0.09 times the mass of the Sun, and this is believed to be about as small as an object can possibly be and still be called a star (this condition will be further discussed later). Groombridge 34 is a binary system composed of two faint red dwarf stars, which was first described in the *Catalogue of Circumpolar Stars* by British astronomer Stephen Groombridge (published posthumously in 1838). The star Lacaille 9352 is

[24] The precise details concerning this remarkable star have only recently been deduced, and are based upon parallax observations made with the Hubble Space Telescope. The details of this study are given by Howard Bond (Space Telescope Science Institute) and co-workers at http://hubblesite.org/pubinfi/pdf/2013/08/pdf.pdf.

so named according to its entry in the (posthumously published) 1763 star catalog constructed by Nicholas de La Caille. Of astrophysical note, this star was one of the first M spectral-type dwarf stars to have its diameter measured – this being accomplished with the Very Large Telescope Interferometer (VLTI) in 2001. Indeed, this low luminosity, red dwarf star has an angular diameter of 1/1,000 of an arc second, and this combined with a parallax of 0.771 mass, indicates a size amounting to just 14 % of that of the Sun. Rather than being a sparkling diamond in the sky, Lacaille 9352 is more of a lustrous ruby.

Lalande 21185, just like Lacaille 9352, is a faint red dwarf star and is named, again, according to its entry in number in an astrometric star catalog – this time the *Histoire Céleste Français* – first published in 1801 (and revised in 1847). This particular catalog was produced at the Paris Observatory under the directorship of Jerôme Lalande.[25] In 1996 Lalande 21185 was elevated from historical obscurity through the announcement, by astronomer George Gatewood (University of Pittsburgh and the Allegheny Observatory), that it harbored at least one and possibly three planets. This conclusion was based upon the careful analysis of the proper motion track of the star as it moved across the sky. Specifically, what Gatewood found was that the proper motion path was not a straight line, as would be expected for a single star, but a complex serpentine trail, and this implied the presence of large, unseen, Jupiter-like planets, in relatively close proximity to the star. The proper motion data for Lalande 21185 appeared to indicate that its planetary companions had orbital periods of about 6 and 30 or more years.

Gatewood's announcement at the 188th meeting of the American Astronomical Society in Madison, Wisconsin, was not the first time that proper motion data had been used to infer the presence of an unseen companion around an apparently single star. Famously, Friedrich Bessel, in 1844, used such observations,

[25] This catalog is historically more infamous for the fact that in 1795 the observers working on its content twice recorded the position of Neptune, but failed to notice that it had moved and, indeed, that it was not a star but a planet. This missed discovery is a little surprising given that William Herschel had serendipitously discovered the planet Uranus just 14 years earlier. The 'official' discovery of Neptune did not take place until it was swept up by Johann Galle at the Berlin Observatory on September 23, 1846.

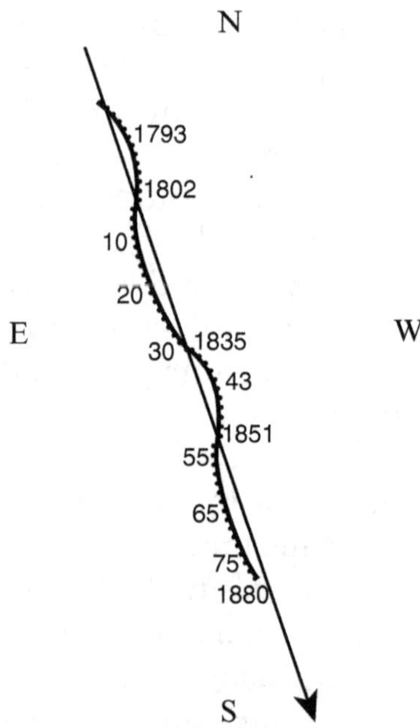

Fig. 1.20 The serpentine, proper motion track of Sirius across the sky from 1793 to 1880. The track is curved, as opposed to being a straight line, because of the presence of its binary companion – Sirius B. The orbital period for the Sirius AB system is 50 years (Image from Camille Flammarion's *Les étoiles et les curiosités du ciel*, 1882)

for example, to deduce the existence of a companion to the bright star Sirius.[26] This, of course, turned out to be the white dwarf companion Sirius B (Fig. 1.20). Likewise, Bessel predicted the existence of a companion to the star Procyon on the basis of its serpentine proper motion track. The faint white dwarf companion Procyon B was later observed for the first time in 1896. Building upon Bessel's example, Dutch-American astronomer Peter van der Kamp (Swathmore College and the Sproul Observatory) later argued, in the 1960s, that Barnard's star had two unseen planets after studying its proper motion path. In this case the data apparently suggested Jupiter-like planets with orbital periods of 12 and 26 years.

[26] Sirius B was first observed by American astronomer Alvan Clark in 1862 while testing the newly constructed 18.5-in. diameter refractor at Dearborn Observatory – then the largest telescope in the world.

Remarkably, with the close of the twentieth century approaching, it appeared that two other stars, besides the Sun, in the solar neighborhood were found to potentially harbor planets. Wretchedly, however, as the first decade of the twenty-first century unfolded, it become increasingly clear that the supposed planets around Barnard's star and Lalande 21185 were nothing more than dreams and shadows; their apparent existence was but the cruel twist of observational uncertainties and perhaps an overly confident interpretation of the limited amount of data available.[27]

Well, such is the history of science, the great triumphs of one era being overwritten by the next. Indeed, it was once suggested that the positional data for the α Centauri AB binary (recall Fig. 1.7) showed small, but regular in time, variations indicative of gravitational perturbations from an otherwise unseen companion. Writing from the Madras Observatory to the editor of the *Monthly Notices of the Royal Astronomical Society*, in a letter dated January 9, 1856, Captain William Stephen Jacob argued that the measured data points "exhibit a very regular epicyclic curve [around] the proper elliptic place." To this observation, Jacob added, "I think, then, there can be no hesitation in pronouncing on the existence of a disturbing body." The "disturbing body" could not possibly have been Proxima Centauri. Its mass is too small and it is too far away from the α Cen AB pair to produce any such effect. Once again, Jacob had found an apparent signal in what was inherently *noisy* data, and his "disturbing body" simply disappeared as more accurate data became available.

There is, of course, a great lesson to be learned from the stories behind the chimera finds of Captain Jacob's, Peter van der Kamp and George Gatewood (and many other researches throughout the history of science). Their efforts are fully laudable, and one should have full faith in the data one has, but to usurp a common saying, extraordinary claims require extraordinary evidence to support them.

---

[27] It is perhaps a little ironic that it was George Gatewood who showed that Peter van der Kamp's analysis of the proper motion shown by Barnard's star didn't support the presence of any planets. More recent studies of both Barnard's star and Lalande 21185, using the Doppler monitoring technique, have failed to find any evidence for the existence of associated planets.

Of the other stars in the solar neighborhood, Sirius is by far the most massive, hottest and most luminous. It is the star of myths and legends, as we have already seen, and it was once, albeit briefly, a standard object of astronomical distance measure. In his great star gauging sweeps of the sky in the late eighteenth century, for example, William Herschel adopted Sirius as a standard brightness star and computed the distances to other stars relative to it. The methodology adopted by Herschel would, in fact, be quite sound if all stars were identical to Sirius – which we now know they are not. It is from Herschel's tentative beginnings, however, that the absolute magnitude and distance modulus method for finding the scale of the stellar realm eventually developed. (See Appendix 1 in this book for details.)

Next to Sirius in brightness, in our sky, is Procyon, so named from the Greek "before the dog," this name following from the fact that it leads Sirius, the Dog Star, across the heavens. While apparently single to the unaided eye, Procyon is, in fact, a binary system. Procyon A is accompanied by a diminutive white dwarf companion (Procyon B) that completes one rotation around the system center every 40 years.

As an F spectral-type star, Procyon A is more massive, larger, hotter and more luminous than the Sun. With an estimated age of three billion years, current theory suggests that Procyon A must be approaching the end of its main sequence lifetime (discussed more fully later) and will soon begin to swell dramatically in size to become a luminous red giant.

Although no planets have been detected in the Procyon system it is presently located just 0.34 pc away from Luyten's star – a remarkably close encounter by galactic standards. Another set of closely situated stars, being about 0.5 pc apart, are YZ Ceti and Tau Ceti. Further ahead in time, in about 32,000 years from the present, it is estimated that UV Ceti will pass within 0.3 pc of ε Indi, making, once again, for a relatively close stellar encounter between stars within the solar neighborhood. Additionally, UV Ceti is the archetype for the class of objects known as flare stars. This group of low mass red dwarf stars, as the name suggests, undergo intermittent and sudden increases in brightness due to the formation of surface flares (analogous to the flares seen on our

Sun). Proxima Centauri is a UV Ceti or flare star, and we shall have much more to say about this in later sections.

Continuing (albeit out of this section's temporal sequence) our forward looking gaze, and greatly magnifying our field of view, the future millennia will not only see the stars of the solar neighborhood complete their various trysts and liaisons, they will also witness a convergence with human history. The year 1977 was remarkable for many reasons. It saw the final eradication of smallpox, the death of Elvis, the incorporation of Apple Computer Inc., optical fibers were first used to carry live telephone messages and the *Voyager 1* and *Voyager 2* reconnaissance craft were launched into space.

Alhough the vast majority of human events occurring in 1977 will likely be of little interest to future historians, the Voyager space probes will endure. They are part of the select group of human-built artifacts now heading into deep space.[28] These diminutive emissaries, made of aluminum, gold and deadly plutonium are now speeding away from the Sun and slowly crawling towards the stars. In about 40,000 years from the present, *Voyager 1* will undergo a stellar encounter with the star Gliese 445, although this encounter will hardly be a lover's kiss, since the closest approach distance will be no more than about a third of a parsec. At about the same time the *Voyager 2* spacecraft will glide past Ross 248 with a stand-off distance of about 0.5 pc. Pressing ever onwards, for another 256,000 years, *Voyager 2* will eventually enter the distant domain contiguous to Sirius, passing by this stellar luminary at a rather frosty closest approach distance of 1.3 pc. By the time that they begin to encounter the stars of the solar neighborhood the Voyager space probes will have long ceased functioning, their instruments having become deep-frozen, dust-impact cratered and mute. Too small to be detected by any conceivable (even alien) technology they will drift almost endlessly through interstellar space bearing silent witness to humanity's first attempts to reach out and touch the stars.

---

[28]The present group of spacecraft heading into deep space are the *Voyager 1* and *2* probes, *Pioneer 10* (launched in 1972) and *Pioneer 11* (launched in 1973). The probe that has traveled furthest from the Sun is *Voyager 1*. This select group of objects will eventually be joined, in the next several decades, by the *New Horizons* spacecraft (launched in 2006) currently on its way towards the dwarf planet Pluto and the Kuiper Belt region beyond.

Shifting our gaze from the distant future back to the past and the near-present we now encounter the stars τ Ceti, ε Eridani and ε Indi. These stars, after the Sun, α Centauri and Proxima, are perhaps the most well known objects within the solar neighborhood. They are Sun-like systems, and they have long held the attention of SETI scientists and science fiction writers. Tau Ceti is a Sun-like star, and it is known to be parent to at least five Earth-like planets, all of which are squashed into a region that could be encompassed by the orbit of Mars in our own Solar System. Epsilon Eridani is a little less massive and somewhat cooler than the Sun, but it, too, may have an accompaniment of planets. Infrared wavelength observations indicate the existence of two asteroid belts, located at 3 and 20 au, respectively, as well as a dust disk in the region 35–100 au from epsilon Eridani. The presence of two distinct asteroid belts is highly suggestive of their being at least two planets within the system – the planets acting to gravitationally shepherd the ring systems and thereby stopping the outer and inner edges of the rings from diffusing outward. Epsilon Indi is again a lower mass star than the Sun, and it, too, harbors a pair of astronomical jewels, making it a triple system and the host to two sub-stellar brown dwarfs.

Not only have τ Ceti and ε Eridani been studied by astronomers to reveal their associated planets, asteroid belts and dust clouds, they have also been surveyed at radio wavelengths in an attempt to eavesdrop on signs of possible extraterrestrial life. Just as the existence of humanity is betrayed to the rest of the solar neighborhood by the chit-chat radio and TV noise leaking from Earth, so similar such signals might betray the presence of extraterrestrial civilizations around other stars. At 12 light years distance from the Sun, the radio signals of Earth's second millennium celebrations have only just swept past τ Ceti. The news that the space shuttle *Columbia* has disintegrated over Texas and that Apple Inc. has opened its on-line music repository iTunes will just be reaching ε Eridani, which is a little closer to us.

Listening in for extraterrestrial radio signals first began at the National Radio Astronomy Observatory at Green Bank in West Virginia in 1960, under the name of Project Ozma and the guidance of astronomer Frank Drake. Similar such surveys have continued, on and off, ever since, but to date τ Ceti and ε Eridani, and

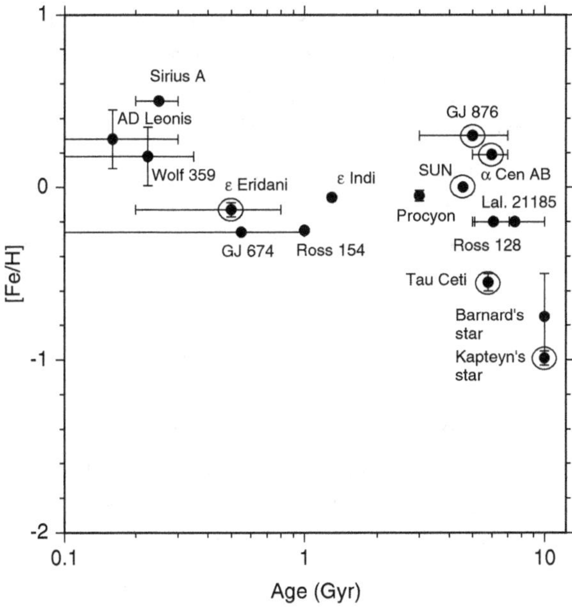

FIG. 1.21 Metallicity versus age for stars in the solar neighborhood. *Circled* data points indicate stars known to be harboring planets and dust disk systems

all the other nearby stars, have remained steadfastly silent. Indeed, the solar neighborhood appears unnervingly quiet, but this voluminous hush only tells us that radio communication is perhaps not that common, and the silence only applies to our present cohorts at the present epoch (more on the SETI later.)

In addition to showing a great range of star types and planetary systems, the solar neighborhood also exhibits a range of system formation times and compositions. If the situation wasn't clear enough before, these two properties, set, of course, by the environment in which the individual stars formed, reveal that the solar neighborhood is a vast multicultural as well as multigenerational assemblage. Figure 1.21 is a diagram in which the metallicity measure of a star is plotted against its estimated age, for those stars within about 15 light years of the Sun. The metallicity measure is based upon the deduced iron (Fe) abundance derived from the spectra of a star normalized to that of the Sun. The larger the metallicity measure for a given star so the more iron-rich it is in comparison to the Sun. Since the metallicity scale is the based upon the logarithm of a ratio, a star with the same iron abundance as the Sun will have a [Fe/H] value of zero; a star with lower iron abundance to that of the Sun will have a negative [Fe/H] value.

We see from Fig. 1.21 that for the stars within about 15 light years of the Sun there is a large range of stellar metallicity values. Barnard's star, for example, is very much lacking in iron when compared to the Sun, whereas Sirius A has much more iron when compared to the Sun. The distinction, of course, is relative, and the values only tell us how the metallicity varies when the Sun is taken as the standard. The fact that there are significant differences in metallicity values, however, is taken as a clear indication that the solar neighborhood at the present epoch is composed of stars that were born in different regions of the galaxy, from interstellar clouds having different iron abundances.

In addition to finding variations in the chemical composition of stars within the solar neighborhood, Fig. 1.21 indicates that there is also a significant variation in their formation ages as well. Again, we see that the estimated age of Barnard's star is over twice that of the Sun and is of order of 10 billion years. The age of the Sirius binary system, in contrast, is estimated to be just a few hundred million years.

Not only do we see a mix of compositions, therefore, but we also see a considerable mix of ages for the stars in the solar neighborhood. There is a slight trend visible in Fig. 1.20, which indicates that younger systems tend to be more metal-rich than older ones, and this partially reflects upon where the star formed within the Milky Way Galaxy. In general, for example, one would expect more metal-rich stars to have formed interior to the Sun's orbit. This situation comes about on the basis of there being more stars within the Sun's orbit than beyond it, and that supernovae, the objects responsible for forming the metals within the interstellar medium in the first place, are more highly concentrated towards the galactic center – the latter effect being a consequence of the Milky Way's star formation history. Interestingly, also, there is a general (although far from universal) trend or preference for planets to form around stars with higher metal abundance. Tau Ceti and Kapteyn's star, however, are the two exceptions, as seen in Fig. 1.21, that break this general rule. In spite of their sub-solar metallicity values, Tau Ceti has five associated planets, while Kapteyn's star has at least two.

This, then, is the solar neighborhood – humanity's present and extended stomping ground. It shifts and it shimmers, the constituent stars streaming past the Solar System like so many parading dignitaries. These stars are the far-flung horizons of future human space adventure, with α Centauri just being the first rung on the unbound cosmic ladder that stretches all the way to the greater galaxy. But for now, we have reached that amorphous boundary of the past sliding into the present, and it is time to hear what modern research has to say about the physical properties of the stars, the Sun and α Centauri.

2. Stellar Properties and the Making of Planets: Theories and Observations

2.1 The Starry Realm

Stars, just like human beings, come in all varieties. They display a multitude of colors, and they are found in densely packed groups or in solitary isolation. They are born, age and die; some living long, quiet lives, others rushing headlong through a luminous youth to an explosive death. The stars, just like human beings, spin and weave their way through space and time; they exhibit spots or flaws of various sizes, they contract and expand, and on occasion thin down and lose mass. Unlike human beings, however, for whom there is no descriptive calculus, the stars are inherently simple physical objects, which is not to say that we fully understand how they form, operate and/or function. In a prescient poem entitled "Mythopeia," dedicated to C. S. Lewis, J. R. R. Tolkien summed up the stellar situation nicely: "A star's a star, some matter in a ball." Indeed, a star is a giant sphere of very hot, mostly hydrogen and helium, gas, and its size is determined according to its age and the manner in which it generates energy within its interior through nuclear fusion reactions.

In this chapter we shall be concerned with the annotation of the stars within the Milky Way Galaxy – their number, their distribution, their physical structure and their relationships one to another. It will be via this extended discussion that the similarities and differences between the Sun and α Cen A and B will be contrasted and compared. Not only this, but the known unknowns, as well as the astronomical issues associated with the α Centauri system in general will be examined. As we shall see, just because α Centauri is the closest star system to us at the present time does not mean that we fully understand it.

M. Beech, *Alpha Centauri: Unveiling the Secrets of Our Nearest Stellar Neighbor*, Astronomers' Universe, DOI 10.1007/978-3-319-09372-7_2,

Let us begin our stellar journey of discovery by first considering the Sun.

2.2 The Sun Is Not a Typical Star

By being such a common, everyday and familiar sight the Sun is often overlooked as a *bone fide* object of astronomical interest. There is perhaps an historical reason for this sentiment, and it should be remembered that it is barely 150 years since it became demonstrably clear, through spectroscopic studies, that the Sun is a star and, up to a point, *visa versa*. Towards the end of the nineteenth century Arthur Searle (Harvard College Observatory) commented in his widely read *Outlines of Astronomy* (published in 1874) that, "Very little, indeed, is known of the stars." He later asserted, however, that, "Observations with the spectroscope have also confirmed the belief previously grounded on the brightness and remoteness of the stars, that they are bodies resembling the Sun." Charles Young further writes, in his 1899 *A Text-Book of General Astronomy*, that "the Sun is simply a star; a hot, self luminous globe of enormous magnitude .... although probably of medium size among its stellar compeers."

With this description, Young confirms the star-like nature of the Sun but has introduced yet another characteristic, stating that the Sun is "probably only of medium size." Accordingly, not only are other stars like the Sun, but there is also a range of stellar sizes, and by implication temperatures, and masses as well. The fact that stars have varying degrees of energy output (that is, luminosity) had already been established[1] about 60 years before Young wrote his text.

Hector Macpherson, in his wonderfully named *The Romance of Modern Astronomy* (published in 1923), picks up on Young's statements by writing that, "The stars are Suns. This is a very good truth which we must bear in mind." Macpherson continues

[1] This fact was evident as soon as the first (believable) stellar parallax measurements were published in 1838/9. Indeed, since the star Vega (as observed by Friedrich Struve) was found to be some 2.2 times further away than 61 Cygni (as observed by Friedrich Bessel) and yet is six magnitudes brighter, it must accordingly have a greater intrinsic luminosity.

to explain, however, that the Sun is a yellow dwarf star. William Benton, in his 1921 *Encyclopedia Britannica* entry concerning the Sun, additionally comments upon its size and notes that, "The Sun is apparently the largest and brightest of the stars visible to the naked eye, but it is actually among the smallest and faintest." The comments by Macpherson and Benton, while in contrast to those of Young, actually build upon the monumentally important results of Ejnar Hertzsprung and Henry Norris Russell, who circa 1910 independently introduced the idea of dwarf and giant stars existing within what is now known as the HR diagram (see Appendix 1 in this book) – a plot of stellar temperature versus luminosity.[2] In terms of stars being blackbody radiators (again a theory not actually established in its modern form until the appearance of the pioneering quantum mechanical model of Max Planck in 1900), the size (radius, R), temperature (T) and luminosity (L) are related according to the famous Stefan-Boltzmann law: $L = \text{constant}\ R^2\ T^4$. That the luminosity is further related to the mass of a star was established by Arthur Eddington in 1924.

By arranging the stars in the HR diagram it is possible to begin comparing the Sun's physical characteristics against those of stars in general. Accordingly, Simon Newcomb, in his *Astronomy for Everybody* (published 1932), explains, albeit rather tentatively, "What we have learned about the Sun presumably applies in a general way to the stars," and that with respect to the HR diagram he notes, "The dot for the Sun, class[3] G0, is in the middle of the diagram."

With Newcomb's latter comment we begin to see a new and quite specific picture of the Sun emerge; it is an average, middle-of-the-road sort of star. Indeed, this comparative point was specifically emphasized by Arthur Eddington in his book *The Nature of the Physical World* (published in 1935). Eddington writes, "Amid this great population [the galaxy] the Sun is a humble unit. It is a very ordinary star about midway in the scale of brilliance.... In mass, in surface temperature, in bulk the Sun belongs to a very common class of stars." To this he later adds (in classic

[2] Here we betray a theoretical bias, since observationally the diagram is a plot of absolute (or apparent) magnitude versus spectral type. The various quantities are, of course, equivalent, but not in any straightforward fashion.

[3] The Sun's spectral type is now described as being G2.

Eddingtonian language), "in the community of stars the Sun corresponds to a respectable middle-class citizen."

With the continued acquisition of data and the development of astrophysical theories, it is reasonably clear that from circa 1930 onwards that the Sun's relative characteristics are generally interpreted as being ordinary or just average. That this notion still prevails within the general astronomical literature is an absolutely remarkable state of affairs since it is patently clear that the Sun is both special, and far from being anything that resembles a typical or ordinary star – it is indeed, extraordinarily special.

The Sun is often described in terms of being typical, average, run-of-the-mill, ordinary, mediocre, and even normal. All such expressions are usually employed in the sense that if a star was picked at random within the galaxy then it would be like the Sun, and/or if one measured a range of values for stellar mass, radius, temperature, and luminosity, then the averages would all somehow reduce to intrinsic solar quantities: 1 $M_{\odot}$, 1 $R_{\odot}$, T ~ 5,800 K, and 1 $L_{\odot}$, respectively.

There are clearly a number of problems with such expectations – not least the fact that this is entirely wrong. When the Sun is described as being an "average" or a "typical" star it is rarely, if ever, stated with respect to what specific distribution of stars. There are, for example, some very obvious comparisons where the Sun would be an extreme and highly untypical object. To the stars in a globular cluster, for example, the Sun would, in comparison, be an extremely young star with a very odd chemical composition (that is, having an extremely high metal abundance). And yet, to the stars in a newly formed OB association, the Sun would by comparison be a low mass, low luminosity, rather old star, with a relatively low metal abundance. Even if we make a more sensible comparison, however, between the Sun's properties and those stars that reside in the solar neighborhood, the Sun in no manner has typical stellar characteristics.

The most complete catalog[4] of stars located close to the Sun with well-measured physical characteristics is that provided by the *Research Consortium On Nearby Stars* (RECONS). Table 2.1 is a summary of the RECONS dataset for the stars located

[4] See the extensive details provided at the Research Consortium On Nearby Stars (RECONS) website: www.recons.org.

TABLE 2.1 Summary of the RECONS data as published for January 1, 2011[a]. The first column indicates the total number of known objects (stars as well as white and brown dwarfs) within 10 pc of the Sun, while the second column indicates the number of stellar systems (single, binary, triple, etc.). Columns three through nine indicate the number of stars of a given spectral type (the Sun, included in the dataset, is a G spectral-type star). Columns 10 and 11 indicate the number of white dwarf (WD) and sub-stellar brown dwarf (BD) objects. The last column indicates the number of planets that have been detected to the present day

Objects	Systems	O	B	A	F	G	K	M	WD	BD	Planets
369	256	0	0	4	6	20	44	247	20	28	16

[a]See the extensive details provided at the Research Consortium On Nearby Stars (RECONS) website: www.recons.org

within 10 pc of the Sun. It is generally true that the vast majority of stellar objects within 10 pc of the Solar System are identified within the RECONS catalog. (This result is probably not true, however, for the brown dwarfs (recall Fig. 1.15), but they do not concern us here.). It is also generally true that the solar neighborhood dataset is representative of that which might be found in any region of the galaxy's disk at the Sun's galactocentric distance of 8,000 pc. A quick glance at the entries in Table 2.1 immediately indicates a predominance of low mass, low temperature, small radii, K and M spectral-type stars. Indeed, the O and B stars are sufficiently rare that the nearest such objects are over 100 pc away from the Sun.

The number of stars of mass M, within the RECONS 10 pc catalog, is described by the mass function $N(M) = 4.6/M^{1.20}$. If there were equal numbers of objects at any given stellar mass then the exponent in the mass function would be zero, but as it stands, of the 320 stars in the 10 pc survey the Sun is among the top 25 most massive. The most massive star within 10 pc of the Sun is Vega, weighing in at just over two times the Sun's mass. The modal, that is, most common, mass value in the 10 pc survey falls in the range between 0.1 and 0.15 $M_{\odot}$, and the median value, for which half of the systems have a greater mass and half have a smaller mass, is 0.35 $M_{\odot}$. That the latter results are more typical for the rest of the Milky Way's disk is revealed by the available data relating to the so-called initial mass function (IMF), which describes the number of stars formed in a specified mass range. Although the slope of the IMF varies in a complex manner according to the mass range being considered, the peak number of stars formed is invariably (even universally) found to fall in the 0.1–0.5 $M_{\odot}$ range.

Sun-like stars having, by definition, a mass near 1 $M_{\odot}$ and thereby a G spectral type are found to make up just 6 % of the stars within the RECONS dataset out to 10 pc. In contrast, the M spectral-type stars constitute 77 % of the total number. Furthermore, the modal absolute magnitude for the stars in the 10 pc dataset is found to be $M_V \approx +13.5$ – a value some 8.5 magnitudes fainter than that of the Sun. Compared to the most typical (that is, ordinary, common, run-of-the-mill, pedestrian, etc.) stars in the solar neighborhood the Sun is nearly 10 times more massive, 10 times larger, 2 times hotter and 10,000 times more luminous. The Sun is not a typical star even within its own precinct.

2.3 How Special Is the Sun?

Given that the Sun is not an average, ordinary or even typical star within the galaxy or the solar neighborhood, is it special in any other way? This question is not intended to focus on humanity's existence – in which sense the Sun is extremely special and we would not exist without it. Rather, the question refers to its defining characteristics such as being a single star, and then a single star with an attendant planetary system, and so on. Again, one can turn to reasonably well known and reasonably well understood datasets to answer this question. Following an approach adopted by astronomer Fred Adams (University of Michigan) the answer to our question can be expressed as a probability.[5] Accordingly, the probability P_{Sun} of finding a star within the galaxy having similar characteristics to the Sun can be written in the form of a Drake-like equation[6]:

$$P_{Sun} = 100\,\mathrm{x} F_1 F_{SB} F_Z F_P F_H \tag{2.1}$$

[5] Adams, F. "The Birth Environment of the Solar System" (*Annual Review of Astronomy and Astrophysics*, **48**, 47, 2010).

[6] The parallel here is to Frank Drake's famous equation for estimating the number of extraterrestrial civilizations within the Milky Way Galaxy. (Frank Drake introduced his now-famous formula for estimating the number of possible extraterrestrial civilizations in 1961, and it has been greatly abused and misunderstood almost ever since.)

The terms entering Eq. 2.1 relate to F_1, the fraction of stars with a mass of order 1 $M_{\odot}$; F_{SB} the fraction of solar mass stars that are single as opposed to being members of a binary or multiple system; F_Z, the fraction of stars with a metal abundance corresponding to that of the Sun at the Sun's location within the galactic disk; F_P, the fraction of solar mass stars harboring planets; and F_H, the fraction of planet-harboring Sun-like stars in which one (or more) might reside within the habitability zone. All of the terms in Eq. 2.1, in contrast to those in Frank Drake's more famous equation, are reasonably well known.

Looking at each quantity in turn, it is evident that $F_1 = 0.06$, corresponding to the fraction of spectral-type G stars within the annotated spectral sequence distribution. To a good approximation $F_{SB} = 1/3$, with the majority of Sun-like stars being found in binary systems (such as in the case of our nearest neighboring system α Centauri AB). F_Z is again reasonably well constrained, and the Sun, in fact, has a relatively high metal abundance, with the survey data indicating that within the solar neighborhood $F_Z = 0.25$ for $Z \geq Z_{\odot}$. Indeed, it should be noted that the composition exhibited by the Sun is not that corresponding to just any radial location within the Milky Way Galaxy, a condition that in fact negates the statements that imply the Sun is somehow situated in an "ordinary" or "nondescript" region of the galaxy. The fraction of Sun-like stars supporting large planets is known to vary with the composition (recall Fig. 1.21) and hence galactic location, and the observations presently suggest that the fraction of Sun-like stars with Jovian planets varies as $F_P = 0.03 \times 10^{Z/Z_{\odot}}$, which is to suggest that $F_P = 0.3$.

And finally, the least well-known quantity in Eq. 2.1 is that relating to the fraction of stars harboring planets within their habitability zone.[7] At present this number may only be constrained via theoretical modeling, but generally it is thought that the fraction of planet-hosting systems harboring habitable planets is something like $F_H = 0.05$.

[7] This term will be defined later on, but its meaning is reasonably clear in that it relates to the zone around a star in which an Earth-like planet might support liquid water (and possibly life) upon its surface.

With our various quantities now in place, the following evaluation is found: $P_{Sun} \approx 0.01$ %. In other words, if one picked a star at random within the disk of our galaxy then there is a 99.99 % chance that it will *not* have the same intrinsic characteristics as our Sun. With such odds against it, clearly, the Sun is not an ordinary star. In addition, the special characteristics associated with the Sun and Solar System apply irrespective of the origins of life on the habitable planet. If we wish to include our own existence in the calculation then P_{Sun} will (according perhaps to one's bias) be many orders of magnitude smaller. Irrespective of this latter addition, by any reasonable standards, the Sun and its attendant planets constitute a rare and uncommon type of system within our galaxy.

If the Sun is a special, decidedly non-typical kind of star within the Milky Way Galaxy, then what is the most typical type of star? The general survey data is, in fact, absolutely clear on this point, and the most typical or most ordinary kind of star that one is most likely to encounter at random within the solar neighborhood (and the greater Milky Way Galaxy) is an object just like Proxima Centauri – a low mass, low temperature, faint, M spectral-type dwarf star.

2.4 There Goes the Neighborhood: By the Numbers

The RECONS data, as summarized in Table 2.1, indicates that the typical spacing between stars within 10 pc of the Sun is about 2.8 pc.[8] That the α Centauri system is located just 1.35 pc away from us, therefore, indicates an unusually close encounter (recall Fig. 1.17).

[8] The typical number of systems (single stars, binary stars and so on) per unit volume of space is 0.09 per cubic parsec. The number of systems in a volume V^* will then be $N^* = 0.09 \times V^*$. If we divide the volume V^* equally between all the stars within its compass, then the volume for each star will be $V_S = V^*/N^* = 1/0.09 = 11.1$ pc^3. The radius r of the sphere having a volume V_S can now determined, and we find $r = 1.4$ pc. Given a typical separation will be of order $S = 2r$, we have a typical system separation in the solar neighborhood of 2.8 pc.

This close proximity is, in fact, even more remarkable if we just concentrate on Sun-like stars. In this case, the survey data reveals a total of 454 Sun-like stars within 25 pc of the Sun, which suggests a typical spacing of 6.5 pc between such stars. Furthermore, the survey data also reveals that only 33 % of Sun-like stars reside within binary systems, which suggests that the nearest twin Sun-like star system to the Sun should, on average, be about 13 pc away. By this standard α Centauri is undergoing an incredibly close flyby of the Solar System.[9] As we shall see later on, Proxima Centauri is an M dwarf flare star,[10] and the spatial density of such stars in the solar neighborhood is 0.056 per cubic parsec, indicating that one might typically expect to find one such star within a sphere of radius 1.6 pc centered on the Sun. On this basis, it can be argued that Proxima is not unusually close to the Solar System. If these same statistics are applied towards α Cen AB, however, then Proxima is remarkably close. Indeed, the odds that Proxima should be located just 15,000 au from α Cen AB purely by random chance are about 1 in 57,000.

Although small probabilities can always be realized the issue of Proxima's close companionship to α Cen AB will be addressed

[9] The twin Sun-like binary star system ζ Retuculi, located just over 12 pc away, is quite possibly the most notorious star system known. In terms of sheer science-fiction horror, it was upon the (imagined) moon Acheron (formally LV-426) orbiting the (imagined) planet Calpamos orbiting ζ^2 Reticuli, that the hapless crew of the mining ship Nostromo first encountered the entirely ruthless, parasitic, jaw-snapping, tail-stabbing, acid-blood-dripping *Alien* (as created by the Swiss artist Hans Giger). Directed by Ridley Scott the movie *Alien* was released to critical acclaim in 1979 and has since spawned a whole number of equally horrifying sequels. The movie prequel *Prometheus* (released in 2012 and once again directed by Ridley Scott) takes the story to another moon (LV-233), and it is revealed that this moon was essentially an abandoned bioweapons installation. On a seemingly more benign front, ζ Reticuli is also associated with the bizarre 1961 abduction case of Betty and Barney Hill. This couple from New Hampshire claims that they were abducted and medically examined by "gray aliens" aboard a landed UFO. Subsequent questioning under hypnosis resulted in Betty Hill recalling a star map that she had been shown. This map apparently revealed 'trade routes" between local star systems, and subsequent analysis by other researchers has linked the home planet of the aliens to ζ Reticuli. Intriguingly, a 100-au radius Kuiper-Belt analog debris disk, possibly hinting at the existence of associated planets, was detected around ζ^2 Reticuli with the Herschel infrared telescope in 2010. The discovery of a Jupiter-mass planet in orbit about around ζ^1 Reticuli was reported in late 1996, but the detection was later retracted and the data explained in terms of stellar pulsations.

[10] Such stars undergo irregular and unpredictable increases in brightness on timescales of minutes to hours.

later on in this chapter. Suffice to say now, however, that it is not entirely clear if it is simply an unlikely random pairing, or a *bona fide* triple star companion.

2.5 That Matter in a Ball

> A star's a star, some matter in a ball
> compelled to courses mathematical
> amid the regimented, cold, Inane,
> where destined atoms are each moment slain
>
> – J. R. R. Tolkien, "Mythopoeia" (1931)

So far we have described stars as being Sun-like, or dwarfs or giants, or of one spectral type or another. Such distinctions are based upon observed characteristics, such as their temperature, their luminosity and their physical size. Indeed, these three parameters describe the position of a star in the HR diagram (see Appendix 1 in this book). Now, however, the question is not so much *what* are the intrinsic characteristics of a specific star but rather, *why* does a star have such and such properties?

By unraveling the orbital characteristics of the two stars within a binary system it is possible to determine their individual masses – literally how much matter they contain. The observations indicate that the smallest stars have a mass of about 0.08 $M_\odot$, while the most massive stars contain about 100 times more matter than the Sun. As we have seen, however, nature tends to favor the formation of low mass stars over massive ones, and the reason for this is entirely due to physics. There is nothing to stop an interstellar cloud collapsing through gravity into an object less massive than 0.08 $M_\odot$, but such an object won't be a star. It will either be a brown dwarf or a massive Jupiter-like object. The reason there is a lower mass limit to *bona fide* stardom relates to the run of internal temperature and density. Below 0.08 $M_\odot$ the central temperature and density of a collapsing gas cloud do not allow for the initiation of internal energy generation through hydrogen fusion reactions (but more on this latter topic in a moment).

Although the lower limit for stardom is set according to the attainment of a minimum central temperature, the upper mass limit is set according to the luminosity (the energy output per second) being too high. In this latter case the problem is not so much that stars more massive than a hundred times that of Sun transgress some forbidden physical limit; it is rather that the in-falling material from the collapsing interstellar cloud can't get to the star's surface to increase its mass. This is an effect related to the radiation pressure built up by the newly forming star becoming so high that it begins to drive any in-falling material outwards again, working against gravity to stop the accretion (and thereby the mass growth) process. Accordingly, the physics of stardom is closely related to energy generation. If the central temperature is too low, then energy generation via nuclear fusion reactions is not sustainable; if the energy release rate is too high, then material accretion is ultimately choked off.

Not only can the masses of the stars within a binary system be determined through the analysis of their orbits, but, with a good distance measurement, so too can their luminosities. If a diagram is constructed in which the logarithm of the luminosity of a star is plotted against the logarithm of its mass, then a remarkable result unfolds. The various data points make up a near perfect straight line. This result indicates that the luminosity of a star is determined by its mass; the more massive a star the greater its luminosity, with the general relationship for low to intermediate mass stars being that $L \sim M^{\eta}$, with $\eta \approx 3.5$.

When this luminosity-mass relationship was first made clear, in the first quarter of the twentieth century, it was realized that, when combined with the HR diagram, it was the mass of a star that dictated its entire appearance. The mass at the end of the star formation process (the moment at which nuclear fusion reactions begin – see below) determines the luminosity of a star. The fact that the star must also first reside on the main sequence (as described in the HR diagram) further dictates that the star must have a very specific temperature (spectral type) and radius. The remarkable mass-luminosity-temperature-radius relationship is illustrated in Fig. 2.1.

The mass-luminosity-temperature-radius diagram for main-sequence stars does not produce a perfect straight line; rather, it shows a small spread in temperatures and luminosities for stars of equal mass. These variations, it turns out, relate to the age of the star (a topic we shall return to later) and its composition – that is, what the star is made of and how much of each specific chemical element it contains. This situation is described according to the

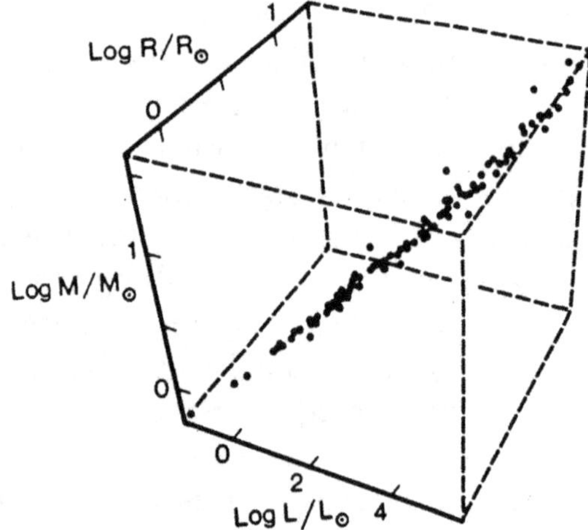

FIG. 2.1 The luminosity – mass – radius relationship for main sequence stars. The data points fall on a diagonal line through the axis cube, rather than being scattered at random (Data from J. Andersen, "Accurate masses and radii of normal stars" (*Astronomy and Astrophysics Review*, **3**, 91, 1991); R. W. Hilditch and A. A. Bell, "On OB-type close binary stars" (*Monthly Notices of the Royal Astronomical Society*, 229, 529, 1987))

so-called Vogt-Russell theorem, which reasons that once the mass and chemical composition of a star are specified, then its internal structure is uniquely determined.[11]

[11] In his classic text, *An Introduction to Stellar Structure* (University of Chicago Press, Chicago, 1939), Chandrasekhar presents a formal definition of the Vogt-Russell theorem: "if the pressure, *P*, the opacity, k, and the rate of generation of energy, e, are functions of the local values of *r* [density], *T* [temperature], and the chemical composition only, then the structure of a star is uniquely determined by the mass and the chemical composition". In many ways the Vogt-Russell theorem isn't a theorem at all. It is essentially a statement about the boundary conditions required to obtain a solution to the collected equations of stellar structure. The theorem has never been mathematically proven, and numerical studies have additionally shown that it is not true under some restrictive circumstances. For newly formed and main sequence stars the theorem is more than likely true and accordingly once the boundary conditions are specified (mass and chemical composition) then a unique solution to the equations of stellar structure will exist. This being said, astrophysicist Richard Stothers (late of the Institute for Space Studies at the Goddard Space Flight Center) found violations of the Vogt-Russell theorem for constant composition, massive stars under certain conditions (see "Violation of the Vogt-Russell theorem for homogeneous nondegenerate stars", *The Astrophysical Journal*, **194**, 699, 1974). Specifically, Stothers found that in the restricted mass range between 170 and 200 solar masses, three envelope solutions, each having different radii, could be 'attached' to a single stellar 'core' solution. Since very few stars form with such high masses, it is probably safe to assume that the non-uniqueness issue is not observationally important. Additionally, present-day numerical models of stars, based upon improved opacity tables and revised in-put physics, do not reproduce Stothers findings.

Without going into details here, stars like the Sun are composed of about 70 % hydrogen, 28 % helium and 2 % all other elements (such as oxygen, carbon, nitrogen, zirconium, and even uranium). What the Vogt-Russell theorem now tells us is that if you change the compositional makeup of a star, then it will take on a slightly different luminosity, temperature and radius; its internal structure will also be somewhat different. The Vogt-Russell theorem also tells us that stars change their observable characteristics (luminosity, temperature and radius) as they age. This result comes about since stars generate their internal energy by transforming one atomic element into another, via nuclear fusion reactions, and this must inevitably change their internal composition. We shall continue the story and implications of stellar evolution in the next section.

Of course, the physics of the situation is a little more complicated than simply describing the mass of a star along with the variation of its internal composition, temperature and energy generation rate. A star, a *bona fide* object, is also an object that continuously hovers on the boundary between collapse, due to gravity, and dispersion, due to the thermal pressure of its hot interior. This condition is known as dynamical equilibrium, and it comes about through a remarkable set of natural feedback mechanisms. The great, if not founding, astrophysicist Arthur Eddington provided a very helpful two-component picture of stellar structure in his famous (but now a little dated) book *The Internal Constitution of the Stars* (first published in 1927). A star, Eddington realized, may be thought of as a material component superimposed upon and continuously interacting with a radiative component. Figure 2.2 illustrates Eddington's basic idea.

The material component in Eddington's picture, as the name suggests, refers to the material out of which the star is made. This is the physical component (the molecules, atoms, ions, electrons and protons) that *feels* the gravitational force, and it is gravity that is trying to make the star as small as possible. The second, radiative component refers to the photons that transport energy in the form of electromagnetic radiation. At the center of the star, where temperatures are at their highest, the photons are in the form of X-ray radiation, but as they progressively move outwards, towards the surface of the star, down the outwardly decreasing temperature

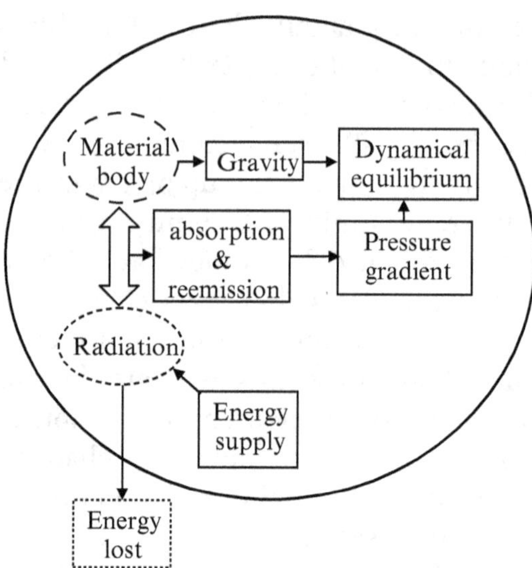

FIG. 2.2 Eddington's two-component star picture. By being hot inside, a star can set up an inwardly increasing pressure gradient that holds the inwardly acting pull of gravity. A dynamical equilibrium is established once the inward and outward forces at each level within a star are in balance

gradient, they continuously interact with the material component, being ceaselessly absorbed, re-emitted and deflected.

Indeed, while the photons travel at the speed of light, and could in theory exit the entire star in a matter of just a few seconds, their journey outwards is slowed dramatically to occupy a timespan of hundreds of thousands of years. Indeed, a photon typically moves just a few millimeters before it interacts with a material particle. By the time that the photons emerge at the surface of the star (from a region appropriately called the photosphere), they are no longer X-rays but light rays, with a characteristic wavelength corresponding to a yellow-orange color.

It is the continuous interaction between their material and radiative components that allows stars to exist. If there was no interaction, the radiation would leak out from the star in a just few seconds, the star would cool dramatically and with insufficient pressure the material component would collapse inward under gravity. In reality a dynamical balance is achieved. By dramatically slowing down the outward journey of the photons the interior of a star can remain hot, and thereby establish an appropriate temperature and pressure gradient at each and every point, to support the weight of overlying material layers. In this manner a star can come into an equilibrium configuration maintaining a constant radius.

The point not so far addressed in this picture is, how does a star remain hot? Clearly stars are losing energy into space at their surface (this is how and why we see them), but if there was no replenishment of that energy then their interiors would eventually cool-off – just as a hot cup of coffee cools off if left standing on a desk. All of this inevitable cooling is encapsulated within the inescapable bite of the second law of thermodynamics. So, to stay hot within their interiors and in balance against gravitational collapse, the stars need an internal energy source, and this is where nuclear fusion comes into play. By converting four protons into a helium nucleus a star can tap a massive internal energy source (literally the hydrogen out of which it is mostly composed), and thereby remain stable for many eons on end. Indeed, we know from the geological record and the study of meteorites that the Sun has been shining (that is, it has clearly not collapsed[12]) for at least 4.56 billion years.

With Eddington's picture in place we can now proceed to describe, albeit briefly, the formation of a Sun-like star. In this description we shall follow a classical approach and consider the pure gravitational collapse of a large, low density, low temperature and spherical interstellar cloud. This picture of collapse will be modified later on when planet formation is discussed.

The starting point of star formation begins with a diffuse cloud of interstellar gas, and we write this symbolically as *Cloud* $(R_{cl}, \rho_{cl}, T_{cl})$ with R_{cl} being the initial radius, ρ_{cl} being the density and T_{cl} being the temperature. The next step is to add in the effect of gravity – and this, of course, will result in the cloud becoming smaller, denser and hotter. The cloud becomes denser since it is envisioned that as time proceeds the same amount of material is contained in a progressively smaller and smaller volume of space.

[12] The dynamical collapse time for the Sun is about 50 minutes.

The temperature of the shrinking cloud increases because as it becomes smaller gravitational energy is released.[13]

As the collapse proceeds the interstellar cloud decreases in size by about a factor of one million, shrinking from an initial cloud radius $R_{cl} \sim 0.1$ pc $\approx 4.4 \times 10^6$ $R_\odot$, to a proto-star size of $R_* \sim 2$–3 $R_\odot$. Likewise the temperature and density inside of the shrinking cloud steadily increase. The end of the gravitational collapse phase is determined by the condition that $T_* > T_{nuc}$ at its center, where T_{nuc} is the temperature at which nuclear fusion reactions can begin. As we shall discuss further below, for hydrogen fusion reactions to begin, T_{nuc} must be of order 5–10 million degrees. Symbolically, the cloud-to-star collapse sequence can be expressed as:

$$Cloud\left(R_{cl}, \rho_{cl}, T_{cl}\right) + gravity \rightarrow Star\left(R_*, \rho^*, T_* = T_{nuc}\right)$$

where it is explicitly taken that $R_* \ll R_{cl}$, $\rho_* \gg \rho_{cl}$, and $T_* = T_{nuc} \gg T_{cl}$.

Why should the gravitational collapse stop simply because nuclear fusion starts? Loosely speaking we can say that the gravitational imperative for continued collapse doesn't go away once nuclear fusion begins; rather it is simply held in check. This is the condition of dynamic equilibrium as described earlier with respect to Eddington's two-component star picture. Turn off the fusion reactions within a star's central core, and gravitational collapse will set in. Indeed, if nothing stops the gravitational collapse, then a black hole will eventually form. At this stage, therefore, the questions we need to ask are, how do fusion reactions work, and how long can they keep gravity in check?

Nuclear fusion reactions, from a star's perspective, are all about the transmutation of one of its internal atomic elements into another. More importantly, however, the stellar alchemy must also proceed exothermically – that is, the process of atomic

---

[13] The reason that a collapsing gas cloud becomes hotter is encapsulated within the so-called Virial theorem. This theorem relates the total kinetic energy K of a self-gravitating gas cloud to its gravitation potential energy U and provides the result that at all times $2K + U = 0$. Since the temperature T of a gas cloud is directly related to the kinetic energy, and the gravitational potential energy is proportional to $-M/R$, where M is the mass of the cloud and R is the radius, so $T \sim 1/R$ since the mass of the cloud is taken to be constant. From this result we see that as the cloud collapses and becomes smaller, so the temperature must become higher. See R. J. Taylor (Note 17 below) for a detailed derivation of the Virial theorem.

alchemy must also liberate energy. It is the energy liberated by the fusion reactions, recall, that keeps the interior of a star hot, thereby enabling dynamical equilibrium to come about.

The essential workings of the energy generation process were first outlined by Eddington in the mid-1920s. It was a wonderful piece of reasoning. Eddington began with the results obtained by chemist Francis Aston, who found that the mass of the helium nucleus, composed of two protons and two neutrons, was smaller by about 0.7 % than the mass of four protons. Here lies the secret of the stellar energy source. Schematically, we have $4P \Rightarrow He - \Delta m$, where 4P indicates the idea of bringing together four protons (hydrogen atom nuclei), He is the helium atom nucleus and Δm is the mass difference indicated by Francis Ashton's laboratory-based measurements.

At this stage the exact details of the fusion reaction process do not concern us. All we need to know is that nature has found a way of taking four protons, converting two of them into neutrons, and then combining the lot in a helium nucleus. The point, as Eddington fully realized, is that if the conversion can be done, then the mass difference Δm is not just vanished away. Rather, using Einstein's famous formula, it is converted into energy, with $E_{4P} = \Delta m\ c^2$, where c is the speed of light. Eddington reasoned, therefore, that while Δm is extremely small per set of 4P conversions the c^2 term is very large, and accordingly only a small fraction of the total quotient of protons within a star need be converted into helium nuclei per second for it to easily replenish the energy lost into space at its surface. To order of magnitude the amount of matter that must be converted into energy per second to power the Sun is simply: (mass → energy per sec.) $c^2 = (E_{4P}/\text{per sec}) = L_\odot$, where $L_\odot = 3.85 \times 10^{26}$ Watts is the Sun's luminosity. This relationship indicates that for the Sun the (mass → energy per sec.) term is about 4×10^9 kg/s – that is, the Sun must convert, through nuclear fusion reactions, about four billion kilogram of matter into energy per second in order for it to shine at its observed luminosity. By human standards four billion kilogram is a lot of matter,[14] but compared to the Sun's total mass of $M_\odot = 1.9891 \times 10^{30}$ kg, the mass lost is

[14] The world production of brown coal and lignite in 2006 amounted to some 1 billion tons, which translates to about 30,000 kg being extracted (on average) per second.

entirely insignificant. Indeed, over the age of the Solar System, a time of some 4.56 billion years, the amount of matter that the Sun has converted into energy is of order 6×10^{26} kg, which is just under 104 Earth masses. This is certainly a large amount of matter, yes, but it is still an insignificant amount compared to $M_\odot$. Indeed, it is just 0.03 % of its mass.

Continuing our order of magnitude calculations, given that the energy liberated per 4P conversion to helium is $\Delta m\ c^2 = (0.007)\ 4m_P\ c^2 \approx 4.2 \times 10^{-14}$ J (where $m_P = 1.6726 \times 10^{-27}$ kg is the mass of the proton), so of order 10^{38} such conversions must be taking place per second in order to power the Sun. That is, the number of protons involved in keeping the Sun shining at any one instance is about 4×10^{38}. Eddington, much to the disdain of book printers, used to like writing out large numbers with all their zeros in place.[15] This certainly emphasizes the sheer scale of the quantities involved. So here goes: $4 \times 10^{38} \equiv$ 400,000,000,000,000,000,000,000,000,000,000,000,000,000. This number can be contrasted against the total number of free protons N_P available to undergo 4P fusion reactions within the Sun, and incredibly, it dwindles thereby into insignificance. We can estimate N_P via the Sun's hydrogen mass fraction, since, indeed, it is the nuclei of the hydrogen atoms that are undergoing the 4P reaction.

Accordingly, $N_P = (0.7)\ M_\odot / m_P \approx 8 \times 10^{56}$. So, once again, only a very small fraction (about 0.000000000000000005 in fact) of the Sun's total number of available protons are involved in generating energy within its interior at any one instant. All in all, it would appear that the Sun can easily power itself by 4P fusion reactions. The question now is, for how long can such fusion reactions proceed?

The nuclear timescale T_{nuc} over which a star can generate internal energy via 4P fusion reactions is estimated by considering how much hydrogen fuel energy it has to begin with divided by the rate at which the hydrogen fuel is used up (or more correctly, converted into helium). Symbolically we have: $T_{nuc} = (0.7)\ 0.007 M^*\ c^2 / L^*$, where M^* and L^* are the mass and luminosity of the star,

[15] The classic example is the Eddington number $N_{Edd} = 136 \times 2^{256}$, which when written out in full is a number 80 digits long. Eddington once commented that he worked out the number long-hand while on a ship crossing the Atlantic.

and the 0.7 accounts for the initial hydrogen mass fraction. For the Sun we find T_{nuc} $(\odot) \approx 2.3 \times 10^{18}$ s (or about 7×10^{10} years).

Detailed numerical calculations indicate that only about 10 % of a star's hydrogen is converted into helium before it is forced to find a new energy source (this topic will be discussed more fully later), and accordingly we have, for the Sun, a nuclear timescale of about ten billion years. Given that the Solar System is already about 4.5 billion years old, the Sun, apparently, is middle-aged, with perhaps another five billion years to go before it evolves into a bilious red giant.

The nuclear timescale formula can be re-cast purely in terms of the mass of a star. To do this we need to recall the luminosity-mass relationship described earlier. Accordingly, we generalize our timescale formula to read: $T_{nuc} = T_{nuc}$ $(\odot)/(M^*/M_\odot)^{2.5}$. This provides us with what at first appears to be a contradictory result. The more massive a star is, so the shorter is its nuclear timescale. This seems odd, at first, since a star more massive than the Sun must surely have more hydrogen fuel. This is true, but the luminosity-mass relationship tells us that as the mass of a star increases so too does its luminosity, and accordingly it uses up its fuel supply much more rapidly. Massive stars live short, but brilliant, lives. In contrast, stars less massive than the Sun lead long, tenebrous lives.

Since α Cen A and B are Sun-like stars, their nuclear timescales will be about the same as that for the Sun – some ten billion years. Proxima Centauri, however, has a mass about 1/10 that of the Sun, and accordingly, it will spend a tremendous amount of time slowly converting its hydrogen fuel supply into energy: T_{nuc} (Proxima) $\approx 2 \times 10^{12}$ years. Incredibly, the nuclear timescale for Proxima is some 169 times longer than the present age of the universe.[16] We shall explore the consequences of the various nuclear timescales relating to the stars in α Centauri in detail in the next section.

The minimum temperature below which the 4P fusion reaction will no longer run efficiently is about ten million Kelvin. Given that the Sun has a central temperature of some 15 million

[16] The universe is estimated to be about 13.7 billion years old.

Kelvin,[17] we can estimate the size of the core region undergoing fusion reactions. First, however, we need an estimate of the temperature gradient within the Sun. This is the measure of how much the temperature drops per meter in moving from the core to the surface. Approximately, the temperature gradient will be $\Delta T/\Delta R = (15 \times 10^6 - 5{,}800)/R_\odot$, where the Sun's surface temperature is taken to be 5,800 K.

With this approximation in place, we find that the temperature decreases only slowly, by some 0.02 K per meter, as we move from its center outwards. The size of the region over which the temperature exceeds ten million Kelvin is therefore $(15 \times 10^6 - 10 \times 10^6)/0.02 = 2.5 \times 10^8 \text{ m} \approx 0.36\ R_\odot$. In other words, the nuclear fusion reactions take place within the inner third of the Sun.

Up to this point we have skirted around the actual physics of the 4P transmutation. Indeed, as Eddington and others were able to do in the 1920s, we can say an awful lot about the inner workings of the stars without knowing the full details of the nuclear fusion process. Eddington once famously quipped of those critical to the idea that stars were not hot enough for the 4P fusion process to take place, that they should, "go and find a hotter place." Indeed, there are no hotter places than the centers of stars in the entire universe.[18]

It turns out, however, that although many fusion reactions are possible, in terms of generating energy from the conversion of four protons into a helium nucleus, stars employ either (or both) the so-called PP chain and the CN cycle. These mechanisms describe the step-by-step interactions needed to complete the transmutation, with each step having its own specific timescale and nuance. The various interaction steps in the proton-proton chain are illustrated in Fig. 2.3, while those in the CN cycle are illustrated in Fig. 2.4.

---

[17] The author has previously provided a series of solutions and approximations to the equations of stellar structure in the book, *Rejuvenating the Sun and Avoiding Other Global Catastrophes* (Springer New York, 2008). See also the highly recommended introductory text by R. J. Taylor, *The stars: Their Structure and Evolution* (CUP, Cambridge, 1994).

[18] Technically the entire universe was hotter than the centers of the stars for a few brief minutes after the Big Bang. But then, at that time, no stars actually existed.

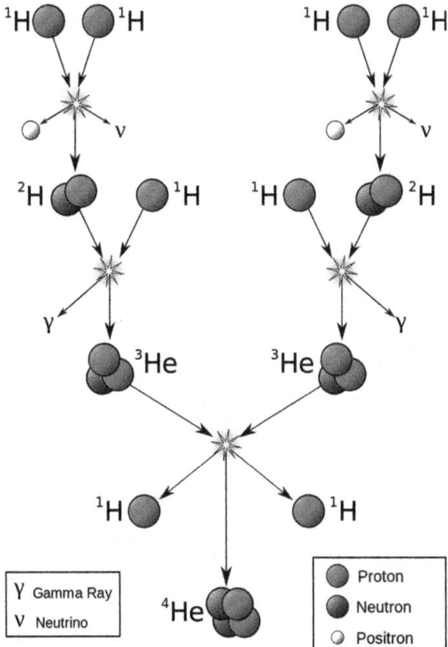

Fig. 2.3 The steps involved in the PP chain. The first step requires the generation of deuterium by the interaction of two protons. This is the slowest step in the entire sequence, since it requires that at the time of interaction one of the protons undergoes an inverse beta decay (to produce a neutron along with a positron and a neutrino). The final step is the interaction of two ^{3}He nuclei to produce a ^{4}He nucleus. A total of six protons are required to produce the two ^{3}He nuclei, but two protons are 'returned' when the ^{4}He nucleus is produced (Image courtesy of Wikipedia commons. FusionintheSun.svg)

Irrespective of which fusion reaction is followed, the PP chain or the CN cycle, the end result is that four protons have been converted into one helium nucleus, and 0.7 % of the mass of the four protons has been released as energy (in the form of gamma rays, neutrinos[19] and positrons[20]) to power the star. The physical conditions under which the two processes can run, however, vary

[19] The neutrinos do not actually interact with the material body of the star and are lost, within a few seconds, into space. In the case of the Sun this neutrino loss turns out to be useful, since by measuring their flux on Earth an experimental test of solar models can be made.

[20] The positrons are an energy source since they will rapidly annihilate with an electron to produce two gamma rays.

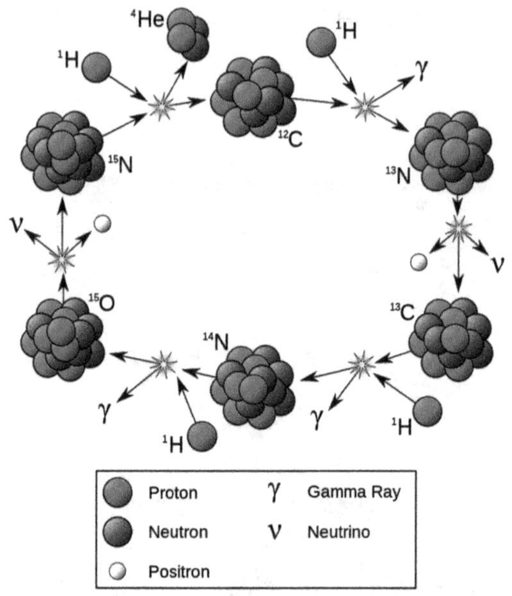

FIG. 2.4 The steps involved in the CN cycle. In this reaction network, the carbon and nitrogen nuclei act purely as catalysts, and the cycle begins with the interaction between a proton and a ^{12}C nucleus. As the cycle precedes nuclei of ^{13}N, ^{13}C, ^{14}N, ^{15}O and ^{15}N are successively produced through beta decays and proton captures (Image courtesy of Wikipedia commons. CNO_Cycle.svg)

and are highly sensitive to both the temperature and density. In general, the detailed numerical models show that the CN cycle operates at higher temperatures and densities than those required for the PP chain. Interestingly, the turnover point at which the total amount of energy generated via the CN cycle begins to dominate over that generated by the PP chain is for those stars just a little bit more massive than the Sun. In fact, with a mass 1.1 times larger than that of the Sun, α Cen A is right on the threshold at which the energy generation mechanism, PP chain versus CN cycle, transition takes place – and this has important consequences for its inner core structure.

So far it has been assumed that the energy generated within the core of a star is transported outwards by the radiative component – the photons. It turns out, however, that energy can also be transported within a star by its material component

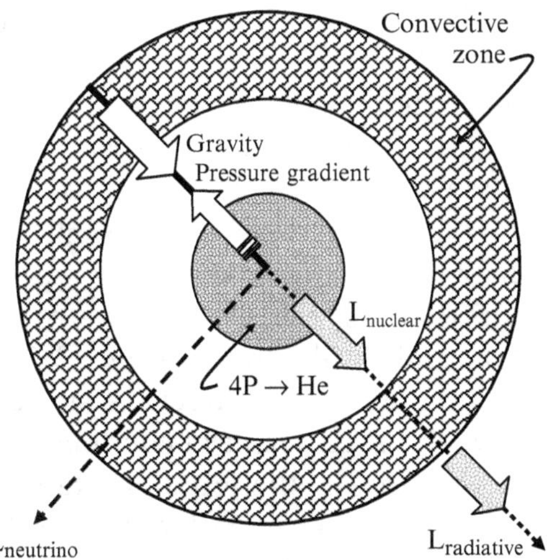

FIG. 2.5 A schematic diagram of the Sun's interior

through convective turnover. In this case a fluid instability literally results in the bulk motion of the material component – just like the bubbling motion seen in a boiling pan of hot water. The mode of energy transport within a star is determined by how much energy there is to be transported outward, the value of the temperature gradient and the ionization state of its constituent material. Detailed calculations indicate that for the Sun, the outer third, by radius, is undergoing convective turnover motion. For α Cen A and stars more massive than about 1.1 $M_{\odot}$, for which the CN cycle begins to dominate the central energy generation process, convective cores begin to develop. For stars less massive than about 0.3 $M_{\odot}$, the entire interior undergoes convective turnover. As we shall discuss shortly, it is the existence of extensive outer convective zones within Sun-like stars that determines their magnetic activity cycles.

Bringing all our results together, we can now construct a schematic diagram of the Sun's inner structure and workings (Fig. 2.5). At each point within the interior there is a dynamical balance between the inward force of gravity and the outward pressure due to the hot interior. The conversion of hydrogen into helium via the PP chain takes place inside the inner third of the Sun's interior, and these fusion reactions generate an outward flow of energy $L_{nucelar}$. The outer third of the Sun's interior undergoes convective motion.

At the photosphere the energy radiated per second into space corresponds to $L_{radiative}$. There is additionally a stream of neutrinos, with a total luminosity of $L_{neutrinos}$, that directly exits from the Sun's core, without any interaction, and streams into space. Given their observed characteristics (see below), the internal structure of both α Cen A and α Cen B will be essentially identical to that derived for the Sun and as illustrated in Fig. 2.5. Humanity may still be many centuries away from directly visiting α Centauri, but we already know, with a high degree of confidence, what the internal structure and workings of the principal stars are like. Hernán Cortés, from that lonely peak in Darien, may well have seen the far-off Pacific horizon and thirsted for adventure (and fortune), but the deeper-penetrating gaze of mathematics and physics has revealed to us the inner workings of the Sun and the far-flung stars. What an incredible result this surely is.

2.6 An Outsider's View

Angel, king of streaming morn, Cherub call'd by Heav'n to shine.

So wrote British poet Reverend Henry Rowe in 1796. Indeed, the Sun, to humanity, is more than just a star; it is our life blood and inspiration. It is also a star that we can see in detail. Indeed, the Sun is one of just a handful of stars that can be resolved beyond a point source into a disk, directly showing thereby a whole host of atmospheric features and phenomena.

It was across the projected disk of the Sun that early telescope-using astronomers, including Galileo Galilei, John Harriot and Christoph Scheiner among others, first observed and traced the motion of sunspots. Against the wisdom of the ancients, the sunspots revealed that the Sun was not a perfect featureless sphere, and moreover, it was not a static sphere. The Sun is spinning, and what is more, later observations by British astronomer Richard Carrington in the 1850s revealed that it was spinning differentially. The time for the Sun to complete one rotation around its equator is some 25 days, while one rotation in the high polar regions takes about 35 days.

The first teasing out of the story encoded within sunspots was begun in the early nineteenth century, and it was started in the hope of finding a new planet. German astronomer Heinrich Schwabe started observing the Sun in 1826, and his intent was to detect the small, dark disk of planet Vulcan while in transit across the Sun. He observed the Sun for over 40 years but never found Vulcan. Indeed, we now know, of course, that there is no such inter-mercurial planet to be found.[21] What Schwabe did find, however, was that the number of sunspots varied in a regular fashion over a period of about 11 years.[22] Schwabe presented his initial observational results in 1843, but the mechanisms underpinning the properties of the sunspot cycle have been challenging astronomers and physicists ever since. The manner in which sunspots are counted and recorded was standardized by Rudolf Wolf in 1848, and it is the time variation of the Wolf number that has been studied ever since.

Working independently of each other Richard Carrington in England and German astronomer Gustave Spörer began studying not only sunspot numbers but also sunspot locations. Although Carrington published first in 1858, the rule describing the variation in sunspot latitude is most commonly called Spörer's law.[23]

Somewhat confusingly, when the data on sunspot latitudes is plotted in diagrammatic form the result is usually called Maunder's butterfly diagram.[24] Moving beyond pure numbers and location, American astronomer George Ellery Hale (MIT) first determined the magnetic nature of sunspots in 1908. Hale's discovery

---

[21] The planet Vulcan was a supposed inter-Mercurial planet. It was estimated to be similar in size to Mercury, but with an orbital radius of about 0.2 au. Many systematic searches for Vulcan were conducted during the later half of nineteenth century – and several observers actually reported finding it! See also Note 44 below.

[22] Schwabe was awarded the Gold Medal of the Royal Astronomical Society in 1857 for his discovery of "the periodicity of the solar spots."

[23] The basics of Spörer's law are this: At the start of each new solar cycle, the sunspots initial appear at mid-latitudes, between 30° and 45°. As the cycle proceeds, however, the sunspots begin to appear at successively lower latitudes. At solar minimum, when the sunspot number is at its lowest count, the sunspots are characteristically found at latitudes ranging between 10° and 25°. At solar maximum, when the sunspot number is at its maximum count, the sunspots characteristically appear within just a few degrees of the Sun's equator. After the time of maximum the cycle begins over again, with the sunspots preferentially appearing at mid-latitudes.

[24] This diagram was first constructed by the husband and wife team of Annie and Edward Maunder in 1904.

followed in the wake of his invention of the spectroheliograph, an instrument that can take an image of the Sun at one specific wavelength of light. With his new instrument Hale found that the spectral lines in regions surrounding sunspots showed the Zeemann splitting effect,[25] and this clearly implicated the presence of strong magnetic fields. Not only were sunspots associated with localized regions of strong magnetic fields in the Sun's photosphere, Hale also found that when sunspots appeared in pairs, they had opposite polarities. Indeed, the magnetic polarity of sunspot pairs shows a 22-year cycle (being twice that of the Wolf number variation and the butterfly diagram).[26] The motion of sunspots not only reveals the differential rotation characteristics of the Sun; it turns out that their very existence also depends upon it. The Sun's magnetic field is generated within its outer third or so by radius through a dynamo process. As shown in

Figure 2.5 the energy transport mechanism in this same outer region is that of convection – literally, the broiling motion of its constituent plasma gas. It is this combination of rotation and convection that combines to produce the Sun's magnetic field and controls the properties of the sunspot cycle. Schematically we have:

$$plasma + rotation + convection \\ + meridianal\, circulation \rightarrow solar\, dynamo$$

Figure 2.6 illustrates the characteristics and operation of the magnetic dynamo. Although the whole process is hugely

---

[25] The so-called Zeeman splitting was first described by Dutch physicist Pieter Zeeman in 1896. Apparently, the story goes, Zeeman disobeyed the direct instructions of his research supervisor and set about studying the effects of magnetic fields on atomic spectral lines. He found that in the presence of a strong magnetic field additional spectral lines could be produced. The first excited state of hydrogen, for example, is split into three energy levels in the presence of a magnetic field; this is opposed to having just one energy level when no magnetic field is present. Though Zeeman was fired for his supervisor-defying efforts, he obtained vindication in 1902 when he received the Nobel Prize in Physics for his discovery.

[26] For the first half of the cycle, for example, the sunspot pairs in the Northern Hemisphere are such that the polarity is *north* for the leading sunspot and *south* for the trailing sunspot (leading and trailing, that is, in the sense of solar rotation). The sunspot pairs in the Southern Hemisphere show the reverse polarity, with *south* leading *north*. This polarity pairing switches during the second half of the cycle, with sunspots in the Northern Hemisphere now having *south* leading *north* polarities, and sunspots in the Southern Hemisphere having *north* leading *south*.

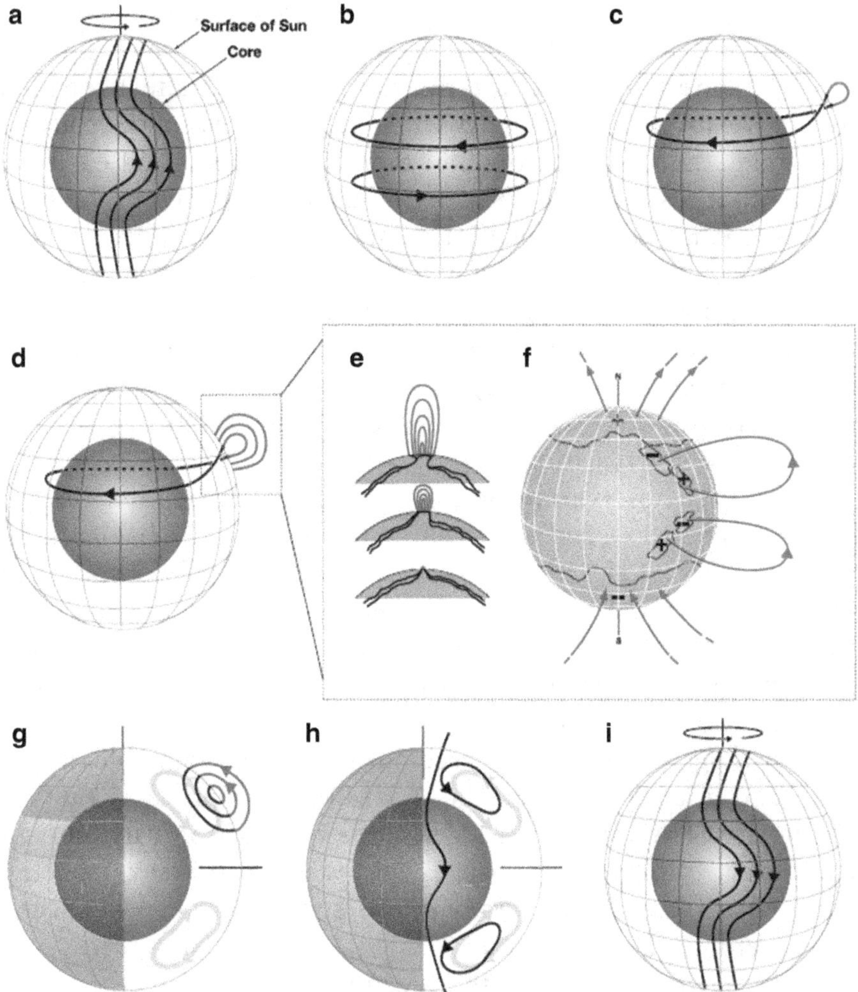

FIG. 2.6 The solar dynamo model. (**a**) The shearing of the poloidal (north–south) magnetic field by differential rotation near the base of the convection zone. (**b**) End result of stage (**a**) and the generation of a toroidal magnetic field. (**c**) Buoyant loops of the toroidal magnetic field rise to the surface, twisting as they do so. Where the loop cuts through the photosphere a pair of sunspots are produced. Further sunspot developmental details are shown in figure (**d**) through to (**f**). Meridional flow (**g**) carries the surface magnetic field poleward, causing polar fields to reverse. Transport of magnetic flux tubes downward to the base of the convection zone at the poles (**h**), resulting in the formation of a new poloidal magnetic field. The newly established poloidal magnetic field (**i**), with the reverse polarity to that in (**a**), begins to be sheared by differential rotation, eventually producing a toroidal magnetic field with the reverse polarity to that shown at stage (**b**) (Image courtesy of Mausumi Dikpate NCAR, Boulder. Used with permission)

complicated, the key principles that are invoked in the operation of the solar dynamo and its accompanying explanation of the sunspot cycle are differential rotation – called the Ω effect – which produces a strong toroidal magnetic field at the base of the convection zone, and then a rising and twisting process – called the α effect – that results in the production of sunspot pairs in the photosphere. It is the meridonal circulation that then stretches and carries the surface magnetic field poleward, establishing the conditions for a poloidal magnetic field and the beginnings of a new magnetic cycle. The basic workings of the $\alpha\Omega$ model and its description of the sunspot cycle were first developed by Horace Babcock in the early 1960s, but the details of the theory are still under active investigation.

The first observation of a solar flare was made by Richard Carrington in 1859, and it was subsequently found that flares are typically associated with active sunspot regions. Indeed, the flares represent the explosive release of magnetic energy, resulting in the generation of a stream of high velocity charged particles and electromagnetic radiation that moves away from the Sun and on into the Solar System.

Although the energy released during a flare is variable, in the more extreme cases it can be a sizable fraction of the Sun's luminosity. The number of solar flares observed is variable and changes according to the sunspot cycle, with perhaps several being observed per day at solar maximum, and maybe one per week being seen at solar minimum.

Although sunspots and flares can be observed directly on the Sun, the overall activity is often gauged according to the so-called S-index related to the strengths of the H and K absorption lines associated with the single ionized calcium atom. This index is high at the times of intense sunspot activity and low at the times when few sunspots are present. The utility of the S-index comes into its own, not so much with the Sun but in the observation of other stars for which the disk cannot be directly resolved. It is a proxy measure therefore for determining magnetic cycle chromospheric activity in other stars.

This method of measuring stellar activity was pioneered by astronomer Olin Wilson at Mount Wilson Observatory in the 1960s. More recently, however, Sally Baliunas and co-workers have reviewed the Mount Wilson S-index survey data and found that 60 % of the stars in the H-K Project survey showed periodic variations, 25 % showed irregular variations and 15 % showed no discernible variation at all (see Fig. 2.7 – Ref.[27]). The magnetic activity cycle of Sun-like stars is apparently variable, and it would appear that such stars can move rapidly from a periodic active phase into one of long-term inactivity and/or high variability. Indeed, it is now clear that the Sun has passed through at least one inactivity phase when the sunspot cycle shut down. Known as the Maunder minimum, after solar researcher Edward Maunder (who first traced its history), it appears that in the time interval between 1645 and 1715 not only were no sunspots or solar flares observed, but the effervescent waves of aurora in Earth's upper atmosphere mysteriously vanished as well.[28]

Although the latter disappearance reveals a link between solar flare activity and upper atmosphere phenomena on Earth, the Maunder minimum, more importantly, coincided with a distinct drop in Earth's global average temperature. When the sunspot cycle stopped, northern Europe lapsed into what is known as the Little Ice Age. It was a time when the river Thames in London would freeze solid each winter and ice fairs could be held across its frozen surface. As the sunspot cycle re-established itself in the 1720s so Earth's global average temperature increased and aurorae were once again seen in the night sky.

The specific mechanisms that produced the Little Ice Age are not fully understood, but the message is clear enough: if the sunspot cycle stops again, and the Mount Wilson Observatory data says that it will, then Earth will face another climate changing challenge.[29] Indeed, the data obtained through the H-K Project at Mount Wilson suggests that on timescales of perhaps thousands of years the Sun should spend of order 20 % of the time in a Maunder minimum-like state. At the present time we have no certain way

[27] The history, current research and rational of the H-K Project at Mount Wilson Observatory is described in detail at: www.mtwilson.edu/hk.

[28] The aurorae are controlled by the solar wind and modulated by solar flare activity. The possibility of a wind of charged particles streaming away from the Sun was first suggested by Ludwig Bermann in 1951.

[29] By saying another we mean in contrast and in addition to the global warming trend, now clearly related to human activity, which is presently forcing Earth's climate towards a rapid and possible devastating change.

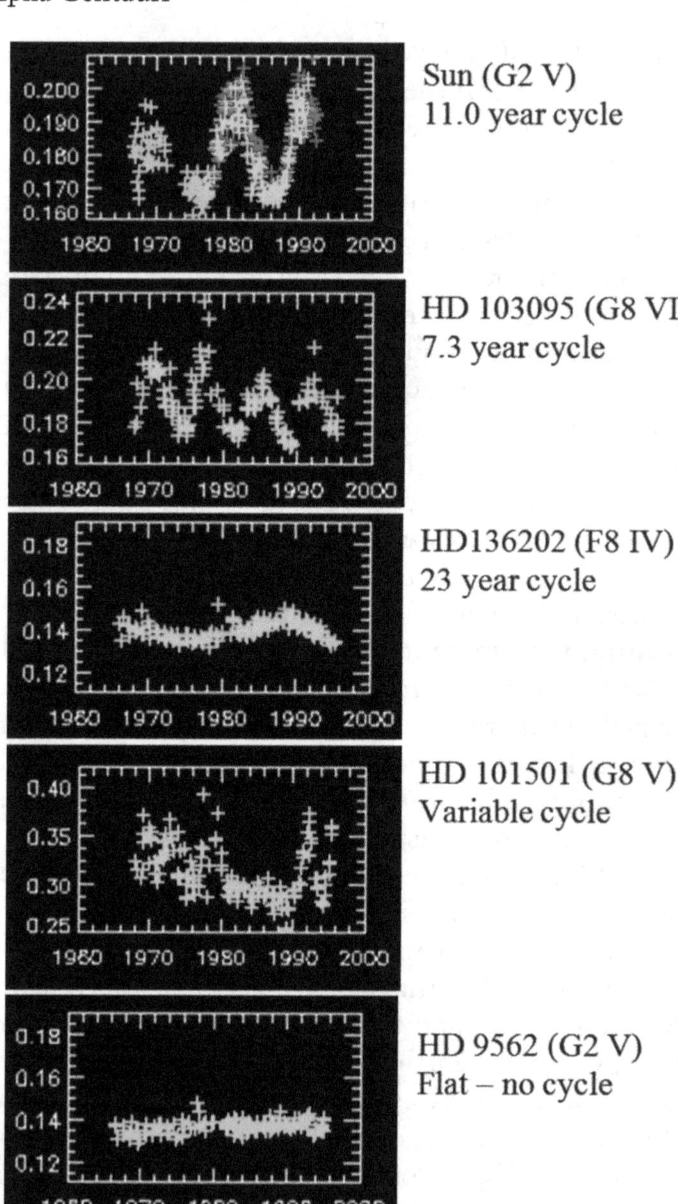

FIG. 2.7 Chromospheric activity of several stars studied in the H-K Project at Mount Wilson Observatory in California. The images show (from *top* to *bottom*) the activity cycle for the Sun, HD 103095 (Argelander's Star), HD 136202, HD 101501 and HD 9562. The activity cycles for the first three stars indicate periods of 10.0, 7.3, and 23 years, respectively. The last two stars show a variable cycle and a flat cycle, respectively (Images courtesy of Mount Wilson Observatory. Used with permission) (The history, current research and rational of the H-K Project at Mount Wilson Observatory is described in detail at: www.mtwilson.edu/hk)

of predicting when the Sun's magnetic activity cycle might switch off again.

Where do α Cen A and B fall with respect to their chromospheric activity? The data appears to be reasonably clear and reveals that α Cen A is in a Maunder minimum-like phase, its activity index having been essentially constant over the past 10 years. This being said, however, Thomas Ayres (University of Colorado) has recently argued that the historical run of data obtained with the ROSAT, XMM-Newton and Chandra X-ray telescopes supports the possibility that α Cen A is either in the process of waking up from a Maunder minimum slumber, or that it exhibits a very long period activity cycle of order 20 years. In contrast α Cen B shows a clear 9-year variation in its chromospheric activity, indicating that its magnetic cycle is a few years shorter than that of the Sun at the present time. Consistent with the study of other Sun-like stars α Cen A and B show a range in their observed magnetic cycle variability. Interestingly, however, as pointed out by Thomas Ayres, near-term future observations of α Cen A may well reveal how the variability cycle picks up again after switching into a deep quiescent mode, and this, of course, may reveal important lessons for us when the Sun once again slides into another Maunder minimum-like phase.

As soon as the means of projecting an image of the Sun's disk onto a screen became available, the blemish of sunspots, along with their variability, was easily noticed. The Sun, however, shows variability in much more subtle ways than the appearance of dark splotches, and indeed, if one looks close enough and in the correct manner its *surface* is found to be pulsing and writhing, with large swaths of the photosphere shifting upwards when other regions are moving down. The Sun is literally ringing, and although there are dominant frequencies the summed effect is a discordant harmony – "like sweet bells jangled, out of tune and harsh."

More than just the circulation of plasma flows within the rising and falling channels of convection cells. This vertical oscillation proceeds through the propagation of acoustic waves. In essence the Sun acts as a resonant cavity for the pressure (that is sound) waves that move through its interior. The existence of these pulsation zones in the Sun's photosphere was first revealed by Robert Leighton (CalTech) and co-workers in 1962. Indeed, by

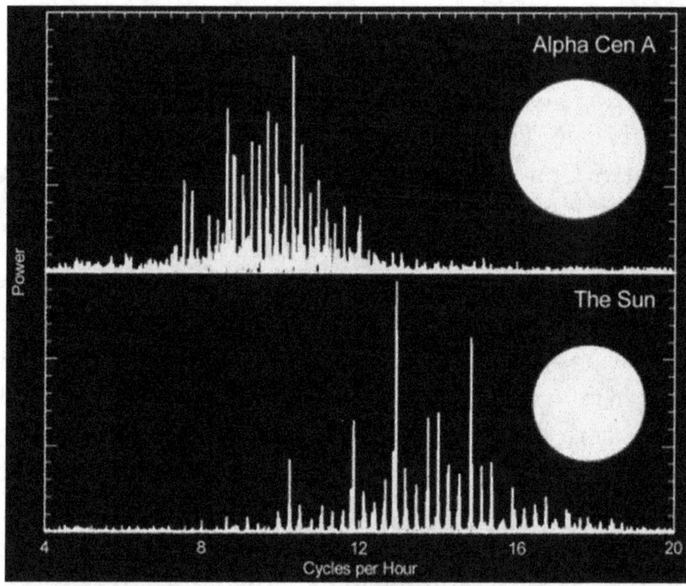

Fig. 2.8 Power spectra for the Sun and α Centauri A. This data reveals the dominant frequencies (where the power is large) of the recorded oscillations. Although many modes of oscillation are present the Sun shows a distinct power spectrum peak close to 12 cycles per h (this is the 5-min oscillation mode). Although α Cen A also shows many oscillation modes, a distinct peak in the power spectrum is revealed at about 10 cycles per h (this corresponds to a 7-min oscillation mode) (Image courtesy of the National Center for Atmospheric Research in Colorado. Used with permission)

studying the Doppler shifts of selected absorption lines Leighton et al. found that localized regions of the Sun's disk showed coherent 5-min oscillations (Fig. 2.8), with the various zones moving either up or down with speeds of order 0.5–1 km/s. From the seeds of helioseismology, literally, the study of Sun-shaking, grew asteroseismology, the study of non-radial star pulsations, and this field of observation now provides some of the strongest constraints upon which to test models of stellar structure.

Asteroseismic studies provide detailed information about stellar interiors, since the observed frequencies of oscillation are directly related to the sound travel time across a star. The speed of sound c in an ideal (perfect) gas is related to the pressure P and density ρ via the relationship $c^2 = \Gamma_1 P/\rho$, where Γ_1 is a constant. By measuring the dominant oscillation frequencies, therefore, a measure of the average ratio of the internal pressure and density can be found, and this can then be compared against the computer model predictions. An additional key point about such studies is that different frequencies of oscillations probe differing depths of a star's interior. Longer wavelength oscillations probe deeper stellar depths than smaller wavelength waves. Not only do the oscillations provide information about the pressure and density of a star's interior, they also provide information about the rotation state of its interior. Such studies, for example, have probed the variation of rotation speed within the Sun's outer convective zone, showing that while the outer regions show differential rotation, the rotation speeds being slower in the polar regions than that at the equator, at the core-envelope boundary (recall Fig. 2.5), the speed becomes uniform. This shows that the core spins like a solid ball. It is in the boundary region of high rotational sheer, the so-called tachocline region, which characterizes the solar dynamo (recall Fig. 2.6, and see below).

Asteroseismic studies of α Cen A and B have been conducted since the early 1980s, with various research groups reporting strong oscillation modes at 7 and 4 min, respectively. Detailed comparisons between theory and oscillation observation have, again, been made by various groups, and these studies have been used to gauge the age of the Centaurian system. Patrick Eggenberger (Observatoire de Geneva, Suisse) and co-workers, for example, used the asteroseismic data to deduce a system age of 6.52 ± 0.3 billion years. Other studies, using differing techniques, have found ages in the range between 5 and 7 billion years for α Centauri, and in general we take the system age of be 6 ± 1 billion years. Compared to the Sun, the stars in the Centauri system are at least 0.5 billion, to perhaps as much as 2.5 billion years older. Not only can the age of the α Centauri system be constrained by asteroseismology but so, too, can their deep interiors.

In this latter respect Michaël Bazot (Universidade do Porto, Portugal) and co-workers have recently reviewed the data relating to α Cen A, and specifically looked to see if there is any evidence that it might have a convective core. As described above one of the conditions under which a convective core might develop in a star is that when energy via the CN-cycle begins to dominate over that of the PP chain – the CN cycle requiring a higher temperatures and core density in order to operate efficiently. The development of such convective cores is important, since they have an effect upon the entire structure and future evolution of a star. Additionally, since there is no fully agreed upon theory to describe convective energy transport within stars, the approximation theory that is used[30] needs careful calibration. It is generally believed that a convective core should develop in main sequence stars more massive than about 1.1 $M_{\odot}$, and accordingly α Cen A sits right at this boundary.

The study conducted by Bazot et al. used a statistical approach to investigate the possible internal makeup of α Cen A. In this manner they constructed nearly 45,000 stellar models, each having slightly different values of the mass, age, composition and mixing length parameter. Comparing this extensive grid of stellar models against the available observations the study revealed an age estimate of about five billion years for α Cen A (this is towards the younger end of the variously published results). The study further revealed a best-fit mixing length parameter of $\alpha = 1.6$, slightly smaller than the value of 1.8 deduced for the Sun.

With respect to the possibility that α Cen A has a convective core Bazot et al. find that the probability is less than 40 %. Indeed, they constrain the core mass and radius to be no larger than 1.5 % and 4 % of the total mass and radius of α Cen A. The situation, at present, remains unclear as to whether α Cen A has a convective core. The odds are not unfavorable, but they are still less than 50–50. Future, higher resolution asteroseismic studies will be required before we can clearly tell what is going on in the core of α Cen A and before we can conduct any similar such parameter study of α Cen B. There are still many secrets that have yet to be unraveled.

2.7 α Cen A and B As Alternate Suns

The stars of α Cen AB are alternate Suns – both literally and physically. The Sun is the prototype, therefore, for understanding their behavior and appearance. Alternatively, the physical properties of

[30] The standard method for describing convective energy transport within a star is the so-called mixing length theory. Here the idea is that a convective *blob* of plasma moves through a specific distance l before dissipating into the surroundings. Generally, the mixing length is specified as being $l = \alpha H_P$, where α is a constant (parameter to be specified) of order one, and H_P is the pressure scale height – the height over which the pressure changes by a factor of $e = 2.71828...$

TABLE 2.2 Physical properties deduced for α Cen A and B compared to those for the Sun

	α Cen A	α Cen B	Sun
Mass ($M_\odot$)	1.105	0.934	1.000
Luminosity ($L_\odot$)	1.519	0.500	1.000
Radius ($R_\odot$)	1.224	0.863	1.000
Temperature (K)	5,790	5,260	5,778
Rotation rate (days)	22.5	36.2	24.5
Composition	$1.5\times Z_\odot$	$1.6\times Z_\odot$	$Z_\odot$
Age (Gyr)	6 ± 1	6 ± 1	4.5
Magnetic field	Yes	Yes	Yes
Magnetic cycle (years)	None (?)	~ 9	11
Oscillations	Yes (7 min)	Yes (4 min)	Yes (5 min)
Planets	??	Yes (?)	Yes

α Cen A and B enable the construction of alternate models for our own Solar System. They provide us with "what might have been" scenarios. Table 2.2 provides a summary of the observationally deduced characteristics of α Cen A and α Cen B and contrasts their data against that derived for the Sun.

The data set displayed in Table 2.2 shows that α Cen A and B bracket the Sun with respect to their mass. They illustrate the dramatic effects that just a 1 % change, plus or minus, in the mass our Sun would have had on the Solar System. For indeed, this small 1 % change in mass, when multiplied through the luminosity-mass relationship, would indicate a 50 % change in the Sun's energy output, and life on Earth would never have evolved. At 1 au from α Cen A the temperature of a Doppelganger Earth would be too hot for liquid water to exist; there would be no oceans, which are the cradle of all life.

Indeed, for the planets as they are in our Solar System, with a central star having the mass and energy output of α Cen A, there would be no habitable planet at all. Mars would certainly be warmer, and it would sit within a region in which liquid water on an Earth mass planet might exist, but its mass at 1/10 that of Earth would still be too small for it to maintain an atmosphere – vital for the safekeeping of oceans – for very long. Exchanging our Sun for α Cen A would result in a lifeless planetary system.[31]

At 1 au from α Cen B the temperature on a Doppelganger Earth would be too low for liquid water to exist; it would be a frozen world sheathed in deep ice. Alternatively, however, Venus (Earth's twin in terms of mass) would now be located within the zone in which liquid water might potentially exist upon an Earth-mass planet's surface. Life, not necessarily as we know it upon Earth, would apparently be possible if the Sun and α Cen B were switched. Once again we learn the important lesson. Earth is a very special place within the universe. The topic of habitability zones, where life on an Earth-like planet might evolve, will be discussed in more detail shortly.

In terms of physical size α Cen A and B are not dramatically different from that of the Sun, being of order 20 % larger and smaller respectively. Their surface temperatures differ only slightly, with α Cen A being just a fraction hotter than the Sun and α Cen B being 500 K cooler. In terms of rotation rates α Cen A appears to be spinning just a little bit slower than the Sun, while α Cen B rotates about 50 % faster.

Detailed spectral analysis of α Cen A and B indicates that for the most part as far as their composition goes they have a similar makeup to the Sun but are definitely richer with respect to many of the heavy elements. Iron, for example, is some two times more abundant in α Cen A than in the Sun. Carbon is only enhanced by a factor of about 1.15, however, and calcium is under abundant by a factor of 0.95. Furthermore, the observations indicate that α Cen B has a slightly higher iron abundance than that determined for α Cen A. Usefully, the generally greater than solar heavy element abundances deduced for both stars in the Centauri system provides us with some insight as to where they might have formed, and it also provides us with the hope that multiple numbers of planets yet await to be found within the system.

[31] The caveat here is that life may still chance to evolve within sub-surface ocean locations such as that found in the interior of Jupiter's moon Europa. In this case the internal heating is provided for by gravitational tidal stretching and as exemplified by the black-smoker ecosystems found in Earth's deepest oceans. Life can find ways to thrive in conditions of complete darkness without the aid of photosynthesis.

That the enhanced heavy element abundances deduced for α Cen A and B is encouraging with respect to the system possibly harboring multiple numbers of planets is based upon exoplanet survey work carried out over the past decade. The data on exoplanet systems and specifically the data on their host stars, indicates that in general planets are more likely to be found the higher the heavy element abundance (Z). Indeed, it appears that the probability increases as approximately the square of the heavy element abundance. Although this probability ostensibly applies to the detection of Jovian, or gas-giant, planets (the actual detection methods will be described later), it is generally believed that the same result will apply to smaller, terrestrial worlds.

The first terrestrial planet in the α Centauri system has already been detected (in orbit about α Cen B – the component with the slightly higher heavy element abundance), and it is probably only a matter of time before more are found not only in α Cen B, but in α Cen A and quite possibly in Proxima as well. We shall pick up this discussion in more detail later.

The idea that the chemical history and evolution of the Milky Way Galaxy is written in the abundances, dynamics and distribution of the stars was first expounded by American astronomer Olin J. Eggen, along with Donald Lynden-Bell (Cambridge University) and Allan Sandage (Carnegie Observatories), in the early 1960s. Accordingly, the stars most depleted in heavy elements are found in the galaxy's outermost halo, moving along highly elliptical orbits with an isotropic distribution around the galactic center. Moving inwards and towards the disk of the galaxy, the stars are richer in heavier elements, and they move in circular orbits around the Sun.

The Sun and α Centauri belong to what is called the thin-disk population of objects, which means that they are relatively young stars moving along circular orbits that carry them no higher than a few parsecs above and below the galactic plane. Not only does the chemical abundance of the stars vary according to the halo and disk structure, the heavy element abundance also increases upon moving closer in towards the galactic center. Specifically, it appears that the history of star formation within our galaxy has favored the inner few thousand parsecs of the disk and core. Since more stars, and importantly, more massive stars, have formed in the inner regions of the galactic disk, so the interstellar medium there is enhanced by heavy elements.[32] Towards the outer boundary of the galactic disk, star formation has been less prolific, and the interstellar medium is accordingly less heavy element-enhanced. That α Cen A and B have heavy element abundances that are somewhat greater than that of the Sun suggests that they probably formed in a region slightly closer-in towards the galactic center – but not by much. Indeed, while it is not possible to say exactly where either the Sun or α Cen A and B (and Proxima) formed (other than within the thin disk component at a radial distance of about 8,000 pc from the galactic center), it is reasonably clear that while they are not common siblings, born of the same natal cloud as the Sun, they are rather distant cousins sired only within the same basic region of the galactic disk.

By comparing detailed numerical models of stellar structure against observed properties it is possible to estimate how old a star might be. In this manner, for the observed mass, temperature and luminosity of star, the compositional abundance terms of a stellar model are adjusted until a good agreement is achieved. Since the internal composition of a star changes systematically with age (as a result of the fusion reactions within its core) so an age can be fixed. The situation is a little better for our Sun, since the laboratory analysis of meteorite fragments enables a formation age to be accurately determined – with the result (as seen before) that the Sun is 4.5 billion years old. When numerical models representing α Cen A and B are adjusted to come into agreement with their observed temperature and luminosity, for their known masses, then ages of order five to seven billion years are derived.

Typically it is taken that the stars of α Centauri are at least some 6 billion years old, making them something like 1.5 billion years older than the Sun. By comparison, therefore, it appears that the Sun is the younger, distant cousin to α Cen A and B. Indeed, a general assessment of star ages in the solar neighborhood finds that the average age is about one billion years older than that of the Sun. It would appear, therefore, that the Sun, the Solar System and humanity are the new(er) kids on the galactic block.

---

[32] It is through supernovae explosions that all of the chemical elements beyond hydrogen and helium are generated and dispersed into the interstellar medium.

Although α Cen A and B are most definitely Sun-like stars, they are not solar twins. Indeed, this latter category of objects is a decidedly select group of objects that not only have the same mass as the Sun but also the same age and composition. At the present time not quite half a dozen stars are known members, or are at least adjunct members, of the solar-twin club.

More solar Doppelgangers are likely to be found in the future, but it turns out that they are relatively few and far between. The closest known member of the solar twin club is the star 18 Scorpii, and it is located at a distance of some 14 pc. Its mass is estimated to be 1.04 ± 0.03 times that of the Sun, and its deduced iron to hydrogen abundance ratio is just 1.1 times higher than that of the Sun.[33] The age estimates for 18 Sco places it between 4 and 5 billion years old – bracketing thereby the 4.5 billion year age deduced for the Sun.

Another solar twin is the star HD 102152, located some 78 pc away. Interestingly for this star, however, is that although it has a near identical mass and composition to the Sun it is estimated to be nearly four billion years older. In essence HD 102152 offers a glimpse of the future Sun.

Although it might seem that 18 Scorpii and HD 102152, given their near perfect solar twin characteristics, are ideal objects to study for possible planetary companions, no new worlds have been located in orbit around them. This, of course, is not to say that none is there, but rather that they haven't been detected yet. Indeed, in the case of these two stars, and for that matter any other solar twin, the most interesting result would be that they are genuinely devoid of planets.

The details of planet formation will be described shortly below, but it is generally taken to be the case that virtually all Sun-like stars should have an associated planetary system. The present paradigm is that low mass stars and planets form in tandem, one with the other and only very rarely separately. Planet-hunting pioneer Geoffrey Marcy (University of California, Berkeley) along with Erik Petigura and co-workers presently interpret the observational situation as indicating that some 26 % of

[33] It additionally has a regular sunspot activity cycle of 7 years duration – similar, indeed, to that of the Sun.

Sun-like stars have associated planets with sizes of between 1 and 2 times that of Earth, with orbital periods between 5 and 100 days.[34] The current observations also indicate that about 11 % of Sun-like stars should have an Earth-like planet located within their habitably zones, with orbital radii between about 0.8 and 1.2 au. Furthermore, Courtney Dressing and David Charbonneau (both of the Harvard-Smithsonian Center for Astrophysics) have also looked at the statistics relating to the low mass, low temperature K and M spectral-type stars, and they find that the occurrence rate of planets with sizes of between 0.5 and 4 times that of Earth, with orbital periods shorter than 50 days, is 0.9 planets per star.[35] In other words, essentially all K and M spectral type stars should have at least one associated planet. To this result can be added the conclusions from another statistical study, of just M dwarf stars, conducted by Mikko Tuomi (University of Hertfordshire, England) and co-workers who find that the occurrence rate of planets less massive than 10 times that of Earth is of order one planet per star.[36]

Given that the present observations imply that all Sun-like and lower mass stars should form with at least one planet, the finding of a genuine planet-less system suggests that some catastrophic processes may occasionally be at play. Indeed, *before*, *during* and *after* planet formation disrupting mechanisms can be identified. The close packing of stars in their natal cloud, for example, leads to a *before* mechanism in the sense that close

---

[34] See, E. A. Petigura et al., "Prevalence of Earth-sized planets orbiting Sun-like stars." This paper can be downloaded at arxiv.org/abs/1311.6806.

[35] See, C. D. Dressing and D. Charbonneau, "The occurrence rate of small planets around small stars" – arxiv.org/abs/1302.1647v2. In addition to estimating the number of planets expected per star, the authors also find that at a 95 % confidence level, the closest transiting, Earth-sized planet located within the habitability zone of its parent star (see Sect. 2.16) should be located within 21 pc of the Sun. Additionally, the nearest non-transiting planet located within its parent star's habitability zone should be closer than 5 pc (16 light years) away (again, at a 95 % confidence level).

[36] See, M. Tuomi et al., "Baysean search for low-mass planets around M dwarfs – Estimates for occurrence rate based on global detectability statistics" – arxiv.org/abs/1403.0430. The results from this study are remarkable since of order 75 % of all stars are red dwarfs. Indeed, the researchers also conclude that perhaps of order 25 % of all M spectral type stars within the solar neighborhood could have super-Earth planets located within their habitability zones (see Sect. 2.16). The data gathered for the study was obtained with the HARPS detector (see Fig. 2.15) and the Ultraviolet and Visual Echelle Spectrograph (UVES) operated by the European Southern Observatory.

encounters between protostars might conceivably destroy their planet-forming disks. The system is then essentially stillborn. A *during* mechanism for planet loss is that of planet migration, where a large Jupiter-mass planet moves inwards and gravitationally scatters any interior planetary bodies prior to interacting with the parent star itself and being consumed via direct accretion. An *after* mechanism would correspond to that of planet stripping via a very close random encounter with another star long after the planets have formed. (See Appendix 2 in this book for the characteristic timescale of such encounter events and also see Fig. 1.17.)

2.8 Proxima Centauri: As Small As They Grow

Nature, for so it would appear, likes to make low mass stars, and Proxima Centauri has about as small a mass that a star can possibly have. Observed as M spectral-type, red dwarfs with low surface temperatures, low luminosities and small sizes, stars like Proxima are located in the very basement of the main sequence. Remove just a shaving of mass from a red dwarf, and it would no longer be a star – rather, it would become a brown dwarf.

Although astronomers are not universally agreed upon an exact definition, it is generally felt that a star is an object that is hot and dense enough within its central regions to initiate hydrogen fusion reactions (recall Fig. 2.3). To achieve these conditions a star, as it forms, must have access to a minimum amount of matter that it can accrete. As before, we can symbolically describe the initial state of a star forming cloud, prior to gravitational collapse, as *Cloud* (R_{cl}, ρ_{cl}, T_{cl}), where R_{cl} is the radius, ρ_{cl} the density and T_{cl} the temperature.

Previously, our argument was that cloud collapse will stop once $T_{cl} = T_{nuc} \approx 10^7$ K, that is, collapse stops once the central temperature is high enough for fusion reactions to begin. With Fig. 2.2 as our guide, it is through the onset of nuclear reactions that a star is able to tap into an internal energy source. The energy generated by the hydrogen fusion reactions then exactly balances the energy lost into space at a star's surface (its observed luminosity). By having a hot interior, a star sets up a pressure gradient, with high pressure at the center and low pressure towards the surface, so that the weight of overlying layers is supported at each point within its interior. The star is then able to find a dynamically stable configuration in which the internal pressure supports the star against continued gravitational collapse.

It is in this manner that at each point within a star the thermal pressure of the interior gas $P_{thermal}$ is exactly balanced by the gravitational pressure $P_{gravity}$ due to the weight of the overlying layers. The thermal pressure is directly related to the density of the gas, assumed at this stage to be a perfect gas in which the individual components do not interact with each other, and the temperature. Working purely in terms of dependent quantities (and ignoring constant terms) we can express the thermal pressure due to the hot interior as $P_{thermal} \sim \rho T$, where ρ is the density of the gas and T is the temperature. The gravitational pressure at the center of a star will be of order $P_{gravity} \sim M^2/R^4$,[37] and when $P_{thermal} = P_{gravity}$ we obtain an approximate expression for the central temperature of $T_C \sim M/R$.[38]

The question we have to address now is, are we sure that the pressure inside of a star can always be described as a perfect gas? And the answer to this is no. Under certain high density low temperature circumstances we may not assume that the gas particles (the atoms, electrons and ions) do not interact with each other. Specifically, the gas within a star can become degenerate, and this dramatically changes the way in which a collapsing gas cloud behaves.

---

[37] The simplest heuristic way to envisage the equilibrium condition is to imagine the star split into two halves, each of mass $M/2$, around its equator. The centers of mass of these two halves, when brought together, will be about a distance R apart, and the area of interaction between the two halves will be πR^2. Using the definition that pressure is the force divided by the area of interaction, and given that our two halves are held together by their mutual gravitational interaction, we obtain $P_{gravity} \approx G(M/2)(M/2)/R^2/\pi R^2$, which gives our result: $P_{gravity} \sim M^2/R^4$.

[38] When $P_{thermal} = P_{gravity}$, we additionally have $\rho T_C \sim M^2/R^4$, and with density varying as M/R^3, we obtain the result that $T_C \sim M/R$. Technically it is the temperature averaged over the entire star mass, T_{av}, that we have just derived, rather than the central temperature T_C. A more detailed derivation gives $T_{av} = 4 \times 10^6 (M/R)$ Kelvin, where now the mass and radius are expressed in solar units. Comparing these results against detailed numerical models we find that for the Sun, $T_C \sim 2.5\ T_{av}$. Additionally, at the Sun's photosphere, $T_{surface} = 5{,}778$ K ~ 10–3 T_{av}.

Degeneracy is a quantum mechanical effect that is related to the Heisenberg uncertainty principle (HUP). This key quantum mechanical principle sets a limit on how well the position Δx and momentum Δp of particle can be known at any one instant. Accordingly, Werner Heisenberg showed in 1927 that $\Delta x\, \Delta p > \hbar/2$, where $\hbar$ is the so-called reduced Planck constant equal to $h/2\pi$. In a degenerate gas, because of the intense crowding, Δx becomes very small, and accordingly the moment Δp must become very large in order to satisfy the HUP. The various particles in a degenerate gas, therefore, must be moving with much higher speeds than would otherwise be expected for a given temperature. Indeed, it turns out that the pressure exerted by a degenerate gas $P_{degenerate}$ is independent of the temperature and only varies according to the density, with $P_{degenerate} \sim \rho^{5/3} \sim M^{5/3}/R^5$.

In the case of the minimum mass for a star to form, the situation is related to which pressure term $P_{thermal}$ or $P_{degenerate}$ comes into equilibrium with $P_{gravity}$ first and thereby halts the collapse. By equating our expressions for $P_{thermal}$ and $P_{degenerate}$ a critical radius $R_{crit} \sim M^{-1/3}$ is revealed, and this provides us (from our earlier expression for the temperature) with a critical temperature $T_{crit} \sim M^{4/3}$. So, in the balance situation where $P_{thermal} \sim P_{degenerate} \sim P_{gravity}$ we have two possible outcomes, depending on the value of T_{crit}. If $T_{crit} > 10^7$ K, then the body can initiate hydrogen fusion reactions before full degeneracy sets in and the body becomes a *bona fide* star with $P_{thermal} = P_{gravity}$. If, on the other hand, $T_{crit} < 10^7$ K then $P_{degenerate} = P_{gravity}$, and it is the degeneracy pressure that stops the gravitational contraction before nuclear reactions can be initiated. Since the degeneracy pressure is independent of the temperature, no matter how much energy the subsequent body radiates into space it will remain stable. A sub-stellar brown dwarf object has accordingly formed. Schematically we now have:

$$Cloud\left(R_{cl}, \rho_{cl}, T_{cl}\right) + gravity \rightarrow$$
$$Brown\,Dwarf\left(R_*, \rho_* = \rho_{degenerate,} T_{crit} < 10^7\,\mathrm{K}\right)$$

Being neither a star nor a Jovian planet, the brown dwarfs form a distinct class of galactic objects. Detailed calculations indicate that the maximum mass for a brown dwarf, which is also the

Table 2.3 Physical properties deduced for Proxima Centauri

	Mass ($M_{\odot}$)	Luminosity ($L_{\odot}$)	Radius ($R_{\odot}$)	Temp. (K)	Rotation rate (days)	Magnetic field	Planets
Proxima	0.123	0.0017	0.145	3,042	25–85	Yes	??

minimum mass for a star, is $M_{limit} = 0.08\ M_{\odot}$, or about 80 times the mass of Jupiter. Of the ten stars nearest to the Solar System, Wolf 359 has the lowest known mass, weighing in at just 0.09 times the mass of the Sun. The star EZ Aquarri C (the 12th closest system to the Sun at a distance of 3.45 pc), has an estimated mass right on the 0.08 $M_{\odot}$ star/brown dwarf divide.

Although brown dwarfs do not initiate hydrogen fusion reactions via the proton-proton chain within their interiors, they can, in their young phases, briefly fuse deuterium via the reaction $D + H \Rightarrow {}^{3}He + energy$. There is again a temperature limit to the onset of these fusion reactions, and detailed calculations indicate a lower mass limit to the brown dwarfs at about 13 times the mass of Jupiter. Objects with masses smaller than the brown dwarf limit are planets. Although the radii of brown dwarfs vary as $R \sim M^{-1/3}$, the radii of planets, which once below the mass of Jupiter tend to have a near constant density, vary as $R \sim M^{1/3}$. An additional distinction between brown dwarfs and planets is planets are thought only to form within the accretion disk surrounding a newly forming star. Planets, in effect, need a parent star to come into existence, while brown dwarfs can undergo a virgin birth through the direct collapse of a small interstellar gas cloud.[39]

Having a mass of 0.123 $M_{\odot}$ Proxima Centauri is about 50 % more massive than the brown dwarf limit of $M_{limit} = 0.08\ M_{\odot}$. So, while Proxima is a low mass stellar object it is nonetheless very much a star, and its variously observed characteristics are summarized in Table 2.3.

[39] Where the population of free-floating Jupiters fits into this scenario has not, as yet, been fully resolved. Although the standard origin scenario for these objects invokes gravitational scattering and ejection after formation within a star's surrounding accretion disk, a recent study by Gösta Gahm (Stockholm University) and co-workers has found evidence to suggest that some may, in fact, be born free through the direct collapse of small "globulettes."

Even though Proxima is already some six billion years old (i.e., the same age as α Cen AB, as described earlier), it has barely started what will be its multi-trillion year stellar journey. For the next many tens of billions of years Proxima's energy output, size and temperature are hardly going to change; it is the quintessential stable star – well, nearly. Though Proxima is in a very stable internal energy generation phase, its outer layers are in erratic turmoil. Proxima is a flare star.

Flare stars were first recognized as a distinct stellar class in the early to mid-1900s. Dutch astronomer Ejnar Hertzsprung serendipitously photographed the very first flare star on the night of January 29, 1924. The unidentified star underwent a sudden and rapid change in brightness for about 1.5 h. Hertzsprung thought that he might have found a new kind of nova. Indeed, it was a nova outburst triggered, he suggested, by the destruction of a small planet in the outer atmosphere of a star. Other stars were soon discovered, however, that showed similar sudden and short-duration outbursts to Hertzsprung's star. Additionally, it was quickly realized that the outbursts were irregular both in their intensity and their duration, and that the time interval between outbursts was entirely random. Not only this, there were just too many repeat outbursts to be the result of planetary in-fall and destruction alone. An internal, rather than an external, mechanism to explain the sudden brightness enhancements was apparently required.

Low mass, red dwarf flare stars are typically classified as being UV Ceti stars – this solar neighbor (just 2.68 pc away – see also Fig. 1.18) being the prototypical star showing irregular flare activity.[40] It is estimated that about 75 % of all red dwarf stars show some form of flare activity, with the outbursts being seen as brightness enhancements across the entire electromagnetic spectrum, from X-rays to radio waves. The flares show a whole range of profile characteristics, but typically there is a rapid rise to maximum brightness followed by a slower decline back to normal. The flares can last from seconds to minutes, and shorter, less energetic flares are more common in occurrence than longer, large energy ones.

---

[40] First described by Dutch astronomer William Jacob Luyten in 1948, UV Ceti is actually a member of a high proper motion binary system (the flare star component is technically identified as Luyten-726-8A).

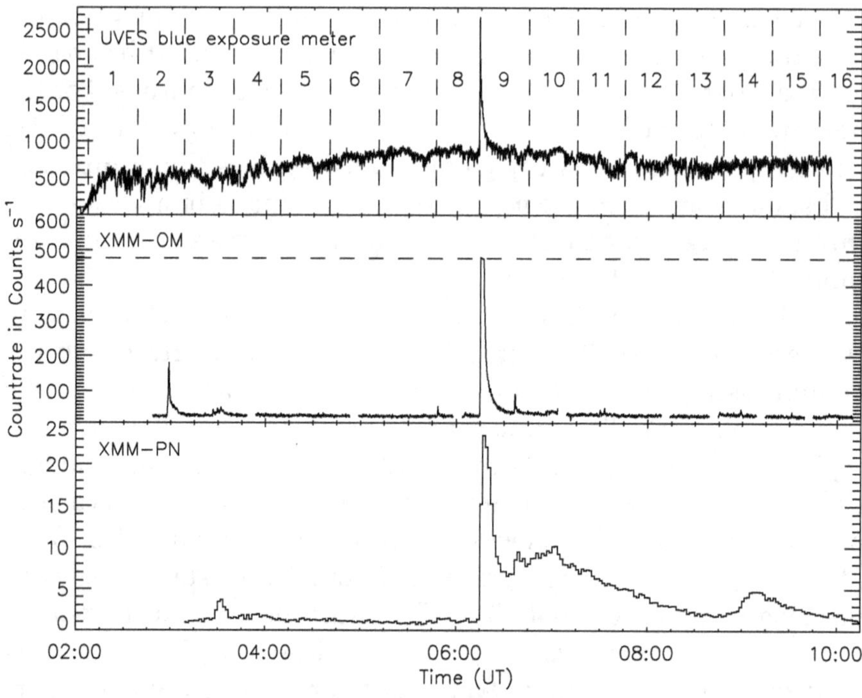

FIG. 2.9 Light curves for Proxima Centauri over a 3.5-h time interval on the night of March 14, 2009. The *top* panel shows optical brightness variations as recorded by the Ultraviolet-Visual Echelle Spectrograph (UVES) attached to the 8.2-m VLT-Kueyen telescope in Chile. The *middle* panel shows the output from the optical monitoring camera of the XMM-Newton spacecraft. The *lower* panel indicates the variation in the X-ray flux as recorded by the XMM-Newton spacecraft. A distinct flare is evident at about 06:15 UT. In the optical part of the spectrum the flare lasts for about 15 min; at X-ray wavelengths the flux is enhanced for nearly 3 h and shows several secondary flare events (Image courtesy of Birgit Fuhrmeister, University of Hamburg. Used with permission)

American astronomer Harlow Shapley, at the time director of Harvard College Observatory, first noticed that Proxima was a flare star in 1951. At that time he commented that, "Dwarf red flare stars may become of considerable importance in considerations of stellar evolution," and in this he was entirely correct. Flares from Proxima have been detected at optical as well as UV and X-ray wavelengths, and Fig. 2.9 shows a number of short duration flares (spikes in the light curves) observed simultaneously from the ground, at the Cerro Paranal Observatory in Chile, and from space with the XMM-Newton X-ray satellite.[41]

Ever since they were first observed the possible mechanisms responsible for producing stellar flares have been a topic of some considerable debate. Although the basic flare mechanism is now understood to be due to the violent release of magnetic field energy, other modulating mechanisms may still be important. These latter processes usually rely upon accretion effects, such as the impact of a comet, asteroid or Kuiper Belt-like object into a star's outer envelope. Indeed, as we shall see later, it is possible that some of Proxima's flare activity is related to its passage through an Oort Cloud structure of cometary nuclei formed around α Cen AB.

Solar flares were first observed on the Sun by Richard Carrington and Richard Hodgson in 1859. From the outset, these localized brightenings were found to be associated with sunspot groups, and accordingly it eventually became clear that they were associated with magnetic field loops. In particular the flares are the result of a process known as magnetic reconnection, in which the magnetic field rapidly rearranges itself, causing thereby a dramatic release of energy. Some of the energy extracted from the magnetic field in a reconnection event heats the surrounding atmospheric plasma, while some additionally goes into accelerating charged particles away from the Sun. In some cases so much energy is released by the Sun's magnetic field that a coronal mass ejection occurs, accelerating massive amounts of material into the greater Solar System.

Such events, if they chance to intercept Earth, result in solar storms and dramatic displays of the aurora. In the case of the Sun, as discussed earlier (Fig. 2.6), the solar magnetic cycle is driven by the αΩ dynamo mechanism. One of the essential components of this magnetic field-generating mechanism is the existence of an inner radiative zone – or more specifically, the tachocline region at the core-envelope boundary. This boundary, located about two-thirds of the way out from the center, is characterized by the presence of a large velocity sheer region. Indeed, it is at this boundary that the rotation changes over from being like that of a solid body to the latitude dependent, differential rotation regime exhibited in

---

[41] B. Fuhrmeister et al., "Multi-wavelength observations of Proxima Centauri" (*Astronomy and Astrophysics*, **534**, id. A133, 2011).

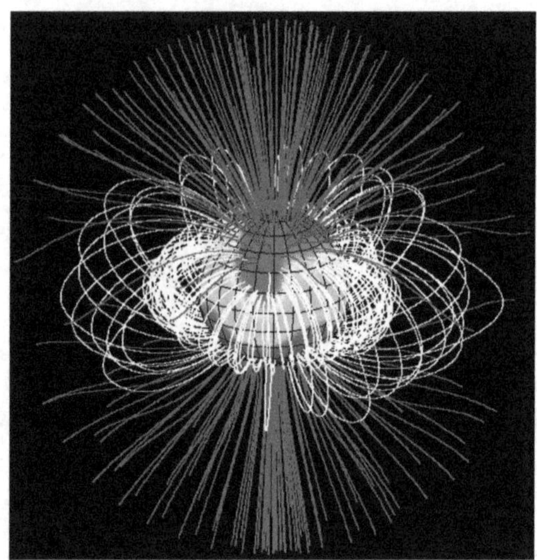

FIG. 2.10 Reconstruction of the magnetic field lines of V374 Pegasi as they extend into space above the star's surface. The topology of the magnetic field is clearly well organized into loops about the equator and polar field lines extending into the surrounding interstellar medium (Image courtesy of M. M. Jardine and J-F Donati. www2.cnrs.fr/en/412.htm. Used with permission)

the convective envelope. It is the characteristics of the tachocline region that determines, in a far from clearly understood manner, the overall properties of the magnetic activity cycle. For Sun-like stars, such as α Cen A and B, there is no specific reason to suppose that the αΩ dynamo mechanism is not at play, and that accordingly it is the mechanism responsible for their observed chromospheric behaviors.

For Proxima, however, we encounter a problem with the αΩ dynamo – the key point being that for Proxima, and indeed all stars less massive than about 0.4 $M_{\odot}$, there is no radiative core. Since such stars are convective throughout their interiors they have no tachocline region within which to anchor a magnetic dynamo, and the question becomes, how can such stars maintain long-lived magnetic fields? For indeed, not only do red dwarf stars have magnetic fields, they also appear to have well-ordered magnetic fields. This latter situation is illustrated by a remarkable study of the M dwarf star V374 Pegasi published in the journal *Science* by Jean-François Donati (Laboratoire d'astrophyhsique de Toulouse et Tarbes) and co-workers in February of 2006 (Fig. 2.10).

Located some 6 pc away V374 Pegasi is about a third the size of the Sun, and detailed modeling of the field line structure suggests that it rotates more like a solid body; this is in direct contrast to the Sun, in which differential rotation dominates in the outer convective zone.

That Proxima, and similar such M-dwarf stars, show magnetic activity is a modern-day mystery and the focus of much detailed research. Indeed, a new mechanism, beyond that of the $\alpha\Omega$ dynamo for generating an organized, self-generating magnetic field, is required to explain why Proxima has a magnetic field and undergoes flare activity.

So, what are the current options? Clearly rotation and convective motion are still going to be important, and the answer to our conundrum has to lie within the physics of these phenomena. One measure that is often used to gauge the extent to which convective motion might be dominated by rotation is that of the Rossby number $\mathrm{Ro} = P/t_{convective}$, where P is the rotation period and $t_{convective} \approx R/\langle Vc \rangle$ is the convective turnover time. R is the star's radius and $\langle Vc \rangle$ is the average velocity of the convective motion. It is known that the Rossby number correlates with chromospheric activity – as described, for example, by the S-index related to the strengths of the H and K absorption lines associated with the single ionized calcium atom. As already indicated the $\alpha\Omega$ dynamo will not operate when the interior of a star is fully convective, but it turns out another mechanism, called the α^2 dynamo, can operate under such conditions, and indeed it becomes efficient once the Rossby number is smaller than about 10. In the α^2 dynamo, the rising and twisting α-effect is the source of both poloidal and toroidal magnetic components. Again, detailed computer simulations indicate that for fully convective stars, in which the α^2 dynamo is at work, a well ordered surface magnetic field can develop (such as observed for V374 Pegasi – Fig. 2.10) even though the magnetic field in the star's interior varies dramatically on many different size scales.

Does the α^2 dynamo work in Proxima? To order of magnitude the convective turnover time is reasonably well known, and with a characteristic convective velocity of $\langle Vc \rangle \approx 5$ m/s we have $t_{convective} \approx R/\langle Vc \rangle \approx 200$ days – which indicates a relatively rapid mixing throughout its interior. The rotation period P for Proxima is not well known, with the variously published measurements suggesting values anywhere from ~25 to ~85 days. Irrespective of the actual rotation period, however, provided it is actually between the currently published estimates, the Rossby number $Ro = P/t_{convective}$ will be much smaller than 10, and this suggests that the α^2 dynamo should be in operation. This result clearly bodes well with respect to explaining why and how it is that Proxima shows relatively strong flare activity.

There is another problem, however, that has as yet to be resolved. One of the outcomes from the numerical simulation of magnetic field generation within fully convective stars is that the surface magnetic field should be constant – that is, there is no modulation mechanism to drive a magnetic activity cycle. And yet, there is every appearance that the flare rate from Proxima is not only variable but cyclic. Using data gathered with the fine guidance sensor on the Hubble Space Telescope, Fritz Benedict (University of Texas at Austin) and co-workers have estimated that Proxima shows an activity cycle of about 1,100 days (~3 years). This variation in activity is further reported by Carolina Cincunegui (Instituto de Astronomía y Física del Espacio, Argentina) and co-workers, but they suggest the period of variation is somewhat smaller and more like 1.5 years. The full situation is still unclear, and exactly what is going on with respect to the observed X-ray emission and chromospheric flare activity of Proxima (and other M dwarf stars) is a challenging and open research question.

2.9 Making Planets

The recipe for making a planet is fairly straightforward and may be easily written down. Understanding the subtle alchemy behind the workings of the recipe, however, continues to be a modern-day research challenge. Using the symbolic formula introduced above to describe the basic star formation process, we need only add one more "ingredient" to begin making planets. Our new recipe proceeds according to the mixing of gravity and rotation:

$$Cloud(R_{cl}, \rho_{cl}, T_{cl}) + gravity + rotation \rightarrow Star(R_*, \rho_*, T_{nuc}) + accretion\,disk$$

By introducing rotation the way in which the interstellar cloud collapses changes from that of a large spherical cloud collapsing radially into a small spherical star to that of a large spherical cloud collapsing into a pancake-like, rotating disk structure. To perhaps overly push our cooking analogy, it is within the pancake that the planets eventually coagulate. The material in the collapsing gas cloud is now envisioned to fall onto the accretion disk and then gradually spiral inward to eventually be accreted by the centrally growing proto-star. The first accretion disk structure to be imaged at optical wavelengths was that associated with the star β Pictoris (Fig. 2.11), and in this case we see the disk edge-on.

Having produced an accretion disk around a newly forming star, a sub-recipe for planet formation must now be introduced. This new mixing procedure operates in such a way that matter clumps begin to form within in the disk – symbolically we have

$$\begin{aligned} &\textit{Accretion disk} \rightarrow \textit{planetesimals} \rightarrow \textit{planets} \\ &+ \textit{dwarf planets} + \textit{comets} + \textit{asteroids} \end{aligned}$$

The key idea of the planet-forming sub-recipe is to turn the gas and dust of the collapsing gas cloud into solid structures of gradually increasing size. Essentially, from the chemistry of the gas and dust grain interactions, molecular structures begin to form. From the molecules new dust-sized grains are produced. From the dust-sized grains, sand grain-sized structures form, and from the sand grain-sized structures, pebble-sized structures accumulate – and so on.

To build a planet, our cooking mantra is, start small and build ever bigger. Not only does solid matter begin to form in the accretion disk, but this recipe in essence cooks itself. Close to the center of the disk, where the proto-star is located, the temperature is high and accordingly only high melting point matter, such as iron and corundum, can exist in the solid phase. Further out the temperature in the disk decreases and so silicates and carbon compounds can begin to appear. Deeper still into the disk the temperature eventually drops to the level at which water-ice can form, and then even further outwards CH_4 and CO ices appear, and so on. The outward decrease in disc temperature drives the chemistry and sorts the basic building materials into specific compositional domains. The important dividing line is that where water-ice can form. The dividing properties either side of the ice line are

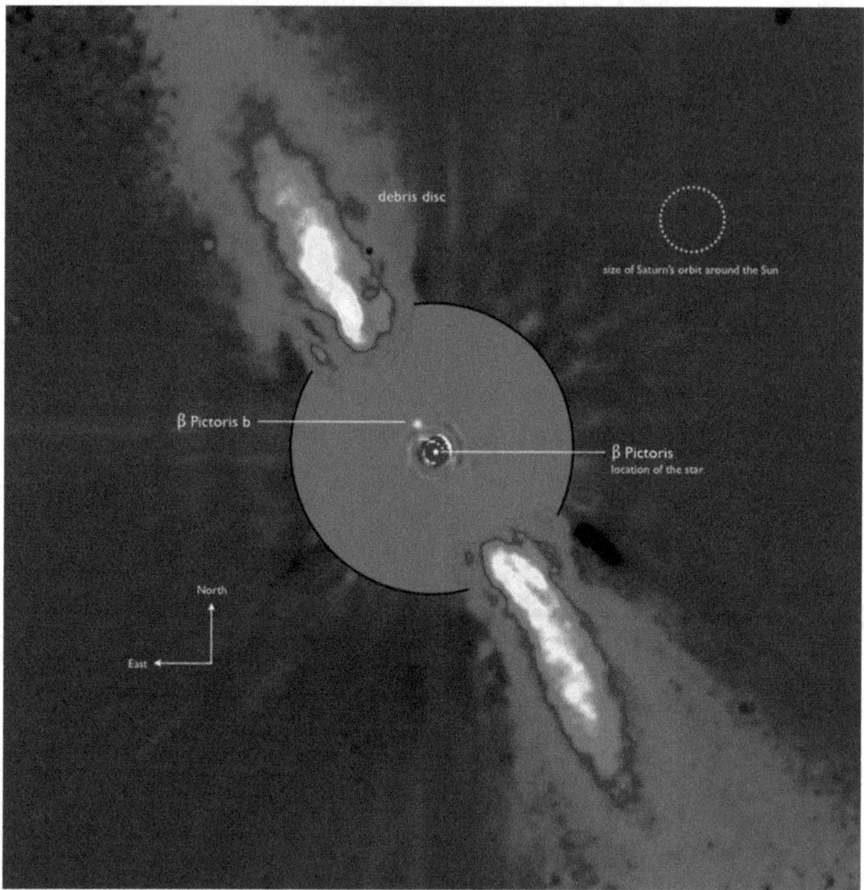

FIG. 2.11 The edge-on disk associated with the star β Pictoris and the planet β Pic b. The star itself has been obscured by an occultation disk, so that the faint light scattered within the disk can be imaged. The disk is about 100 au across, and at least one Jupiter-mass planet has formed within it. The circle to the upper right indicates the scale according to the orbit of Saturn (19 au across) in our Solar System (Image courtesy of HST/NASA)

distinguished in our Solar System according to the characteristics of the terrestrial and Jovian planets. Inside of the ice line, which for Sun-like stars is located some 3 au into the disk, terrestrial planets, made predominantly of silicates and iron, form. Beyond the ice line, the massive Jupiter-like planets grow.

Although the temperature and ice line determine the basic compositional makeup of the disk, the planets themselves are built-up by random collisions – a literal hit and stick process. The first kilometer-sized structures to appear in the disk are called planetesimals, and it is through the collision and accretion of these objects that planets are eventually produced. In the Solar System the leftover planetesimals, not actually accreted into a planet, are observed as cometary nuclei and asteroids.

The processes of collision and accretion, collision and break apart continues within the disk until a few gravitationally dominant structures appear. These will ultimately be the planets. Having formed, however, the process of orbital sorting is far from over, and the observation of exoplanets clearly informs us that migration, especially of massive Jovian planets, is common. Indeed, by migrating inwards, from beyond the ice line where they were formed, the hot Jupiter planets are produced. As part of this migration inward, planet-on-planet gravitational interactions and scattering will additionally take place, and this will result in the ejection and possibly orbit flipping of interior planets (i.e., the terrestrial planets that formed interior to the ice line). The inward migration and gravitational scattering process is also the most likely mechanism for producing cold Jupiters. These are represented by the Jovian exoplanets located at many tens to even hundreds of au from their parent stars.

The formation of planets around stars with binary systems is not greatly different to that for single stars. The only caveat relates to how close the two stars in the system approach one another. Again, it is the mutual gravity and tidal forces between the two stars and their individual disks that will determine the outcome of planet formation. Detailed numerical simulations of the accretion growth process show that a close companion can either enhance the planet formation process or it can totally destroy it. Several research groups have specifically studied the formation of planets in α Cen AB, and the general consensus is that there is no specific reason to suppose that planets cannot form there. The real questions are: where have the planets formed, and how many planets are there?

Although some of the details will be discussed below shortly, it appears unlikely from both the observations and the planet-formation modeling studies that either α Cen A or B has any associated large Jupiter-mass planets. Part of the reasoning behind this conclusion is that the ice line for these stars will be located at about 2–3 au, and this is very close to the limit set for stable orbits (discussed further below). Additionally, it has also been suggested that disk-disk gravitational interactions in the newly forming α Cen AB system might act to suppress giant planet formation and favor the production of close-in terrestrial planets. There is no present consensus on the exact details, indicating of course that we could easily be surprised by what is eventually found, but the numerical simulations suggest that planets in the mass range from sub-Earth to perhaps 1–2 times the mass of Earth may exist about both α Cen A and B with orbital radii between about 0.5 and 2.5 au. Theoretically it would appear that we are good to go. There is no specific physical reason to suppose that planets cannot exist within the α Cen AB binary, and the challenge now is to see if any such objects can be found observationally.

2.10 New Planets and Exoworlds

In a strange way the response of both the media and the public to the discovery of the first planet in the α Centauri system was rather muted. Certainly the discovery and initial announcement made the headlines, but within just a few days the whole show was over and seemingly done with. We have indeed become a jaded society, overwhelmed and inundated by tabloid gossip and trivial pursuits. Perhaps the lackluster response was a Northern Hemisphere effect. After all, α Centauri is not visible from Russia, most of China, Asia, Europe and North America, countries where the greater part of the world's overburdened population lives. Indeed, an informal poll reported in the *Huffington Post* for October 17, 2012 (one day after the planet's discovery was announced) found that only 54 % of the people interviewed in San Francisco had heard of α Centauri, and less than 1 % of those asked knew that it was the nearest star system. Perhaps the stilted public response was because some 850 other exoplanets had been discovered before α Cen Bb was identified – just another distant world in a long (and continuously growing) list of un-seeable external worlds, another planet whose features can, at the present time, only be imagined rather than experienced through direct imaging.

Well, in spite of this subdued response, the discovery of α Cen Bb was a scientific triumph – a triumph of observational technique, hard work and of detailed system analysis. Indeed, the discovery of α Cen Bb was the result of some 20 years' worth of human perseverance, intellectual tenacity and technological development.

There is no clear beginning to the story of planet and exoplanet discovery. Certainly, philosophers have been speculating upon and astronomers actually looking for additional planets within our own Solar System, and around other stars, for a very long time. Perhaps, stretching the point at issue a little, the Greek philosopher Philolaus (c. 470–385 B.C.) might be credited with creating the first new planet within the universe. As a member of the Pythagorean School, Philolaus held the number 10 in great esteem. It was the tetraktys, the holy or mystic number. In applying this numerical reasoning to the universe, however, Philolaus realized that there was a problem. He knew there were eight 'planetary zones' – which corresponded to the regions of Mercury, Venus, the Sun, the Moon,[42] Earth, Mars, Jupiter and Saturn. And he knew there was a zone for the stars (encompassing the celestial sphere), making in total a nine region dichotomy of the heavens. However this division, Philolaus argued, did not resonate with the importance of the tetraktys, and therefore he speculated that another planet, the counter Earth, must exist.

To satisfy the ideal of Pythagorean numerical harmony, Philolaus reasoned a whole new world into existence. With history repeating itself, the same manner of philosophical thinking once again appeared, some 2,000 years after Philolaus, to bring into existence the planet Neptune (discovered in 1846). In this latter case, however, a much greater power of numerical calculus and logic was employed to argue that a planet must exist – specifically it was required to explain the observed residuals in the motion of Uranus.[43] As always, however, nature loves to toy with human hubris, and the same philosophy that resulted in the successful detection of planet Neptune failed in the case of planet

[42] Both the Moon and the Sun, recall, were viewed as planets in the classical era.

[43] It was by working through the prohibitively complicated mathematics describing the mutual gravitational interaction that would result between Uranus and a hypothetical perturbing planet that led Urbain Joseph Le Verrier and John Couch Adams to successfully predict the properties of the perturbing planet's orbit.

Vulcan – an imagined world postulated to explain the observed motion of planet Mercury.[44]

The eventual discovery of Uranus was inevitable; but as luck would have it the person who saw it as something other than a star was William Herschel. Other observers had recorded Uranus's position on star charts long before Herschel made his results known, but they failed to recognize it as a new world. Indeed, Herschel first thought that he had discovered a new comet, and it was only later he realized he had actually discovered a new Jovian-type planet.

Herschel may well have been fortunate in his planetary discovery of 1781, but he greatly enhanced his chances of success through the very act of pursuing a thorough and systematic study of the heavens. When Herschel began his star gauges it was really just a matter of time before Uranus would swim into his view. Furthermore, there was every reason to believe that additional planets might well exist beyond Saturn (located 9.5 au from the Sun) since Edmund Halley had demonstrated that at least one periodic comet, Halley's Comet, moved as far as 35 au away from the Sun during its 75-year-long orbital sojourn. Indeed, when Halley made his famous prediction in 1707, later confirmed in 1758, his comet (when located at aphelion) more than trebled the size of the then known Solar System.

With the discovery of planet Uranus something extraordinary happened. A new, apparent harmony emerged for the description of planetary orbits. The result would probably have pleased Philolaus and his fellow Pythagoreans, but it continues to trouble astronomers to this very day. This controversial new harmony relates to the so-called Titius-Bode law that was written down and willfully copied by various authors during the mid- to latter part of the eighteenth century. It is a simple mathematical rule that says

---

[44]When Le Verrier tried to explain the anomalous motion of planet Mercury he invoked the same idea that had resulted in the successful finding of planet Neptune. To this end a new inter-Mercurian planet, given the name Vulcan, was postulated. Planet Vulcan, however, was later written out of existence by the equations of general relativity developed by Albert Einstein in 1916. Indeed, Einstein showed that the observed anomalies of Mercury's orbit were entirely due to the Sun's curvature of spacetime. The story of Vulcan is further described in the author's book, *The Pendulum Paradigm – Variations on a Theme and the Measure of Heaven and Earth* (Brown Walker Press, Florida. 2014).

that the orbital radius a of each successive planet within the Solar System is given by the relationship: $a(\text{au}) = 0.4 + 0.3 \times 2^m$, where $m = -\infty$, 0, 1, 2, 3, ... and so on. The sequence for m is certainly odd, starting as it does with a negative infinity that suddenly jumps to a value of zero and thereafter increases by a factor of one in each successive step, but for all of this, it does provide a remarkably accurate expression for the observed orbital radii of the planets in the Solar System – up to a point, that is.

For the planets Mercury ($m = -\infty$) through to Saturn ($m = 5$), the comparison between the formula result and the observations is shockingly accurate. Further pushing the boundaries of credulity the law, for $m = 6$, also describes the size of the orbital radius for planet Uranus. Seemingly, this law has great predictive powers, and astronomers soon argued that the apparent gap in the planetary system at $m = 3$, corresponding to $a(\text{au}) = 2.8$, must contain some undiscovered object.

Sure enough, on January 1, 1801, Giuseppe Piazzi swept up the first of the asteroids. Ceres, as this new object was to be named, is the largest object in the main Asteroid Belt between Mars and Jupiter, and it has an observed orbital radius of 2.7654 au. In many ways the results were, or more to the point are, entirely unreasonable. Why should such a simple mathematical expression as encompassed within the Titius-Bode law provide such an accurate description of planetary orbits? As we saw earlier, the formation of planets is a random, dynamic, and chaotic collision- and accretion-dominated process, and there is no underlying reason to suppose that such complex stochastic processes can be explained by a mathematical rule based on one simple variable and three simple constants. And yet, this appears to be what nature has given us – up to a point.

In spite of its remarkable accuracy in describing the orbital radii from Mercury out to Uranus, the Titius-Bode law fails horribly with respect to its predictions for the orbital radii of Neptune ($m = 7$) and Pluto ($m = 8$). Indeed, for Pluto the formula is in error by more than 100 %. Clearly, there is more to the construction of the Solar System than the dictates of the Titius-Bode law. University of Toronto researchers Wayne Hayes and Scott Tremaine demonstrated this latter point in a wonderful 1998 publication in which they showed that Titius-Bode-like laws could be constructed for almost any random configuration of stable planetary orbits. Hayes and Tremaine also found that the best fit Titius-Bode law for the entire Solar System is: $a(\text{au}) = 0.450 + 0.132 \times (2.032)^{n}$, $n = 0, 1, 2, 3, \ldots, 8$. This new law removes the strange (if not highly suspect) $-\infty$ first power for Mercury, but it now no longer shows any satisfying numerical elegance in its form. The new constants jar the eye.

Well, of course, beauty isn't everything, but it would appear that at best Titius-Bode-like laws are nothing more than useful numerical coincidences that come about due to the fact that if a planetary system is going to remain stable over long intervals of time, 4.56 billion years in the case of the Solar System, then planetary spacings had better satisfy some basic physical principles. Indeed a kind of Goldilocks rule is likely to apply, with the planets not being too close together, else gravitational perturbations will ruin the orbital stability, and yet not too far apart either, since it would appear that if the basic building blocks are in place then nature will build a planet if it can – in other words large gaps in planetary systems are unlikely.[45] Additionally, the planets within the Solar System appear to favor orbits in which the orbital periods of each successive pair satisfies a near mean-motion resonance. In this manner, Mercury orbits the Sun (approximately) five times for every two orbits of Venus (this is a 5:2 mean motion resonance[46]); Venus and Earth exhibit a 13:8 mean motion resonance. Likewise,

[45] On purely geometrical grounds, ignoring gravitational interactions, one can argue that in order to avoid collisions any pair of planets must be arranged so that the aphelion distance of the innermost planet must not be further away from the Sun than the perihelion distance of the outermost planet. This condition can be cast in terms of the orbital periods of the two planets such that Pout/Pin > 1, where the out and in subscripts indicate the inner and outermost planets respectively. Using Kepler's third law this result can be case in terms of the semi-major axis of each planet's orbit so that, $P_{out}/P_{in} = (a_{out}/a_{in})3/2$. Excluding the pairing between Jupiter and Mars, the typical value for P_{out}/P_{in} in the solar system is observed to be about 2. Using this result, we obtain for the non-overlapping orbits condition that $a_{out}/a_{in} \sim 1.6$. We can now, in fact, use this condition to 'predict' the existence of the asteroid belt between Mars and Jupiter. For Mars, $a_{in} = 1.5$ au, so in keeping with the other planetary pairings within the solar system, we might predict the presence of a planet at $a_{out} = 1.5 \times 1.6 = 2.4$ au, and this is indeed just about where the asteroid belt begins – it is also comparable to the orbital radius of the dwarf planet Ceres ($a = 2.77$ au). Yet another 'planet' could be squeezed-in before we reach Jupiter at aout = $2.4 \times 1.6 = 3.84$ au. A planet interior to Mercury might also be predicted upon the non-overlapping orbits condition, and in this case $a_{in} = 0.246$ au. Of historical interest the orbital semi-major axis of the latter 'planet' corresponds to that predicted by Le Verrier for Vulcan (see Note 44).

[46] Saturn and Jupiter also exhibit a near 5:2 mean motion resonance, while Neptune and Pluto exhibit a strict 3:2 mean motion resonance.

since orbital stability requires the avoidance of very close approaches between successive pairs of planets so the development of near circular orbits with regular spacings is favored, with the spacing being modified according to the various masses of adjacent planets.

We now see the Titius-Bode law not as some profound physical statement but as an underlying shadow framework for describing planetary spacings within a stable planetary system. There is indeed every reason to suppose that all multiple exoplanetary systems that are stable over long intervals of term will obey some form of a Titius-Bode-like law; strangely, however, its universality lies within the fact that it is simply an ordered sequence of numbers and not a fundamental physical law describing the formation of planetary systems. Remarkably, therefore, it does appear that the Titius-Bode law has the power to predict the existence of planets, but its power is analogous to a trick performed by a well-trained magician rather than a result derived by a reasoned astrophysicist.

The next obvious question becomes, therefore, "Do exoplanetary systems obey Titius-Bode-like laws and can we use them to find otherwise unobserved planets?" The answer to this question is, as we shall see below, yes; but before we can further discuss the issues some details on how exoplanets are detected should be put in place.

2.11 Planets Beyond

The idea that planets might orbit other stars is far from being a new one. Indeed, it is an ancient idea. The atomistic philosophy of Epicurus (341–270 B.C.) supposed, in fact, that there were an infinite number of stars and planets, and specifically an infinite number of Earths. Much later in history, the scripturally misguided polymath Giordano Bruno (1548–1600) reasoned that not only did it make philosophical sense that the universe was infinite in extent, but that every star in the universe should also have an attendant planetary system. René Descartes (1596–1650) further argued, half-a-century after Bruno's condemnation and execution, that the universe was filled with circular eddies in which matter could accumulate. Furthermore, at the center of each vortex, Descartes reasoned, a star would eventually form, and each newly birthed star would have an associated set of sibling planets.

Three-hundred and fifty years further on from Descartes, we now know that the universe isn't infinite in extent, although it is certainly large and relatively old (being brought into existence some 13.8 billion years ago), and it certainly contains many stars. There are something like 10^{23} (100,000 billion billion) stars in the observable universe. Remarkably, however, although the physics behind Descartes vortices has been entirely discredited, and while Bruno had no supporting evidence for his other worlds idea, they were both right in asserting that virtually all low mass and Sun-like stars will have attendant planets. Indeed, modern astronomers suggest that finding a Sun-like star without attendant planets is the oddity, rather than the other way around.

Titius-Bode law guidance aside, all the new, that is non-classical, planets within the Solar System have been found telescopically. In this manner the new discoveries timeline has progressed mostly as a result of technological advancements – bigger telescopes and more sensitive detectors enabling astronomers to find smaller, fainter and more distant worlds. There is a limit to this process, however, and after a while the basic point-and-look approach will no longer yield new discoveries. In order to find exoplanets it turns out that a kind of peripheral vision needs to be applied. Astronomers don't actually look for exoplanets directly, but rather they look for the effect of such planets upon their parent stars – either via astrometric measurements, the Doppler effect or through repeated brightness transients.

We briefly described the astrometric method earlier. Here the presence of a planet is revealed by mapping out the path of the parent star across the sky. Such observations are non-trivial, and highly time consuming. In essence, however, with the astrometric technique one is trying to separate out a sum of motions: the star's proper motion, the star's parallax and the star's reflex motion due to its planetary companion. Ignoring (or more precisely, correcting for) the six monthly parallax variation in position, the reflex motion combines with that of the star's proper motion to produce, over many years, a serpentine path across the sky (recall Fig. 1.20 for Sirius). If there was no planetary companion, and hence no

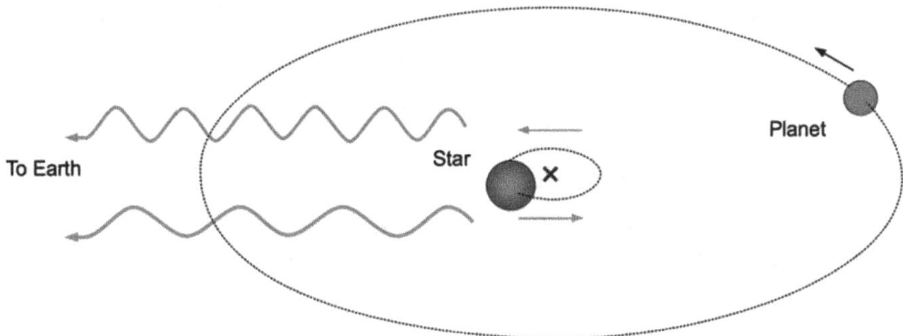

FIG. 2.12 The Doppler method of planetary detection. The unseen planet induces a reflex motion of the star around the system's barycenter (marked *X*), and this motion can be quantified by monitoring the variations in the star's radial velocity, as measured through its photospheric absorption lines, over time. It is the periodic blueshift (motion towards) followed by redshift (motion away) variations in the radial velocity measurements of the star that betray the gravitational presence of the planet (Image courtesy of Wikimedia commons. Radial_Velocity_Exoplanet.png)

reflex motion, then the proper motion path would be a straight line across the sky. The serpentine motion comes about because the star and planet move around a common center of mass (or barycenter) that is displaced away from the center of the star, and because the proper motion actually tracks the straight line motion of the barycenter through space. The star's radius of motion about the center of mass is given by $a_S = a_P(M_{planet}/M_{star})$, where a_P is the planet's radius of motion. The more massive the planet and the larger a_P, so the larger is the reflex displacement of the star. Astrometry, therefore, is all about measuring the displacement a_S. As discussed earlier, the discovery of planets via astrometric techniques has historically proved ineffective, but this is primarily because it comes into its own when looking for large mass (that is brown and/or red dwarf) companions, when a_S is relatively large.

The Doppler method (for details see Appendix 2 of this book) of exoplanet detection also relies upon the measurement of a reflex motion, but in contrast to the astrometric method it operates best when a_S is small (see Fig. 2.12). The reflex motion again comes about because the system's center of motion is displaced away from the center of the parent star. It is a remarkable celestial dance that takes place, with the existence of invisible worlds being betrayed through the barely measurable do-si-do that is stepped out by the apparently single parent star. By directly measuring, over many days, months, years and even decades, the velocity with which the parent star moves about the system's barycenter it is possible to deduce the masses and orbital periods of its associated planets. Indeed, in the ideal case, where the planet has a circular orbit and when we are fortunate enough to see the orbit edge-on (this maximizes the Doppler shift signal), then the system of equations to solve for are:

$$\left.\begin{aligned} V_S &= \frac{2\pi\, a_S}{P} \\ M_{\text{star}} &= \frac{a^3}{P^2} \\ M_{star} a_S &= M_{planet} a_P \end{aligned}\right\} \tag{2.2}$$

where V_S is determined via the Doppler shift variations, P is the orbital period (again measured from the radial velocity variations) and $a = a_S + a_P$. In the second relationship shown in Eq. 2.2, which is actually Kepler's third law of planetary motion, it is assumed that the mass of the star is very much greater than the mass of the planet. To fully determine the orbital radius a_P and mass of the planet M_{planet}, an appropriate value for M_{star} must be specified. By algebraically combining the equations listed in Eq. 2.2 it is possible to show that

$$V_S^2 = 4\pi^2 \left(\frac{M_{planet}^2}{M_{star}} \right) \left(\frac{1}{a_P} \right) \tag{2.3}$$

where the typical case in which $a_S \ll a_P$ is assumed.

From Eq. 2.3 we now discover an important distinction between the astrometric and Doppler techniques for finding planets. Although the astrometric technique works best for companions with large orbital radii (large a_P values), the Doppler technique works best, that is produces a larger and more easily measured velocity signal, when the planet's orbital radius a_P is small – that is, close in towards the parent star. Conversely, Eq. 2.3 indicates that the smaller the planet mass and the greater the distance it is from the parent star, so the smaller is the velocity variation signal.

The idea that planets might be detected in orbit about distant stars through the Doppler monitoring of reflex motions was first discussed in the 1950s, but it was not until the early 1990s that the observational techniques were in place to make such studies feasible. The technical challenge that planet detection presented was that the velocities to be measured were in the range of meters per second, rather than the kilometers per second that astronomical spectroscopes had otherwise worked to. In the case of the Sun, for example, the reflex velocity induced by Jupiter amounts to a 13 m/s variation (Fig. 2.13). The radial velocity induced by Earth is about 0.1 m/s. Not only is the velocity small, but for an extraterrestrial civilization monitoring the Sun, they would have to take measurements over at least 12 years, the orbital period of Jupiter, before it was clear that a planet had actually been detected. Exoplanet hunting, if our Solar System is taken as typical, is not for the hasty or faint of heart. Luckily for astronomers, however, it now appears that our Solar System is not typical, and that the existence of planets around other stars can, on occasion, be the subject of just a few weeks worth of (hard and exacting) work.

The discovery of the very first exoplanet was announced in the august pages of the journal *Nature* for November 23, 1995. The authors of this historical work were Michel Mayor and Didier Queloz, astronomers working at the Geneva Observatory in Switzerland. It was a remarkable piece of work, with a remarkable and entirely unexpected outcome. The two observers had embarked upon a spectroscopic survey of Sun-like stars in early 1994, and after some 18 months of data collection had identified a number of candidate stars that showed the promise of having attendant planets. The system that they specifically chose to concentrate upon, however, was 51 Pegasi, a Sun-like star located some 15.4 pc from the Solar System.

Mayor and Queloz explain in their research paper that the first observations of 51 Peg were obtained in September of 1994, and that by January 1995 the first indications of a short-period planetary companion were evident – a result that was later confirmed during two dedicated observational campaigns in July and September of 1995. The radial velocity variations of 51 Pegasi were undoubtedly periodic, alternately showing redshifts and blueshifts of 60 m/s (see Fig. 2.14). A new world, 51 Peg b, had been

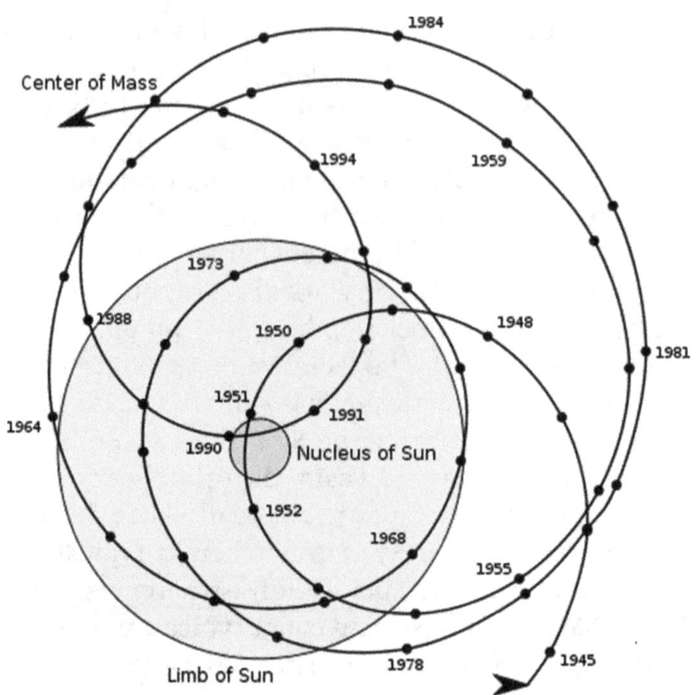

FIG. 2.13 The reflex motion of the Solar System's barycenter due to motion of the planets around the Sun (Image courtesy of Wikimedia Commons. Solar_system_barycenter.svg)

discovered, and the radial velocity data indicated a planet having a mass about half that of Jupiter moving on a close-in orbit with respect to its parent star.

Incredibly, the new planet had an orbital period of just 4.23 days and moved along a near circular orbit with a radius of just 0.0527 au. This result was unprecedented, and a good deal of initial doubt and pessimism had to be overcome before all astronomers agreed that a new planet had, in fact, been detected. The problem, as described earlier, was that no theory in the mid-1990s predicted that gas-giant planets might be found any closer than about 3 au from a Sun-like star. Having an orbital radius nearly 100 times smaller than the expected lower limit at which Jovian planets should form clearly required further investigation, but Mayor and Queloz confidently asserted that the problem of 51 Peg b lay with the theory and not with the observations – and they were, of course, entirely right.

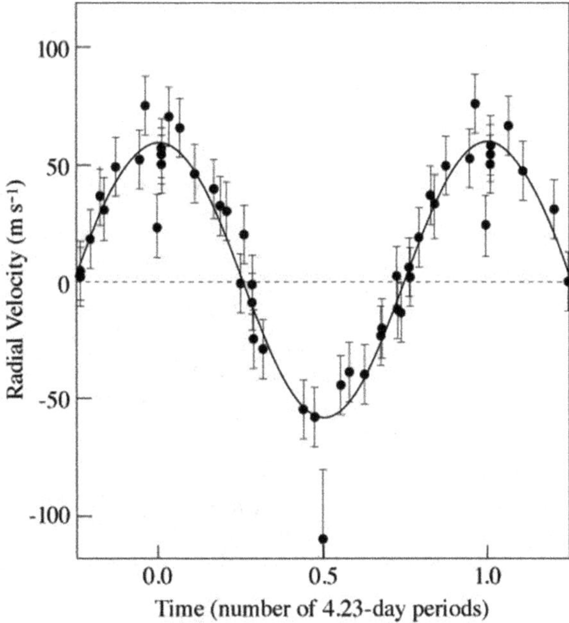

FIG. 2.14 The regular radial velocity variations of the star 51 Pegasus, indicating the presence of an attendant planet – 51 Peg b. The observed 54.9 m/s maximum radial velocity and the 4.23 day period indicate that 51 Peg b has a mass of 0.45 $M_{Jupiter}$ and an orbital radius of 0.0527 au (Diagram courtesy of NASA's Cosmos andTufts University)

Numerous research groups have developed extremely sensitive techniques for measuring exoplanet Dopper shifts, but the state-of-the-art system at the present time is HARPS – High-Accuracy Radial Velocity Planetary Searcher). Developed by the European Southern Observatory (ESO) consortium, with Michel Mayor as principle investigator, HARPS saw first light in 2003 and is attached to the 3.6-m telescope at La Silla Observatory in Chile. The HARPS system (Fig. 2.15) is all about stability and precision. The central component is a ruled grating that splits the incoming starlight into a very high resolution spectrum. The star spectrum is simultaneously compared against a thorium-argon calibration spectra, which not only allows for a very precise evaluation of the stellar absorption line wavelengths (the critical part of the radial velocity measure), but it also allows for extremely precise instrumental drift corrections.

FIG. 2.15 The HARPS spectrograph, shown here with its vacuum chamber casing open. The heart of the spectrograph is the rectangular echelle diffraction grating (seen slightly *above* image center) (Image courtesy of ESO)

Indeed, to help improve instrument stability not only from day to day but from year to year, the whole instrument is housed within a large vacuum vessel in a temperature-controlled environment. Such attention to detail has enabled HARPS to provide long-term radial velocity measurements to an accuracy of 1 m per second, and since operations began it has assisted in the discovery of more than 150 exoplanets. A second instrument, HARPS-N (the N standing for Northern Hemisphere) has recently been commissioned and housed upon the 3.58-m Telescopio Nazionale Galileo Telescope on La Palma; this instrument saw first light in 2012. The HARPS-N instrument has been highly successful in helping to characterize a number of the transiting exoplanets discovered by the Kepler spacecraft (see later).

In the 15 years since the discovery of 51 Peg b, 1,791 additional exoplanets have been discovered around some 1,111 stars (as of May 27, 2014). Planets, indeed, appear to be almost everywhere; they orbit single stars, they orbit binary stars, and they roam freely through space.

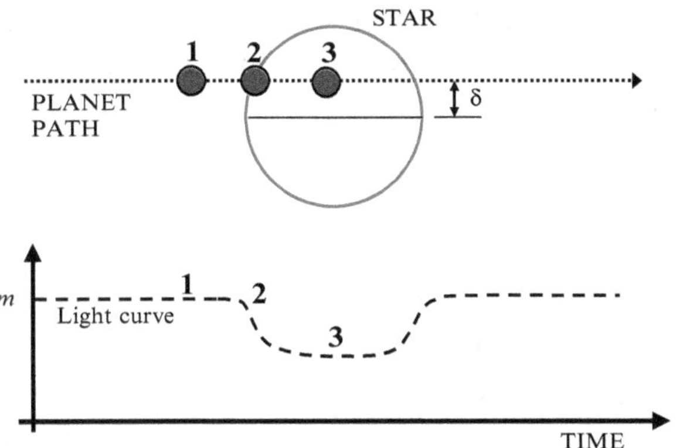

Fig. 2.16 The transit method of planet detection. The light curve, brightness versus time, diagram, for a star hosting a planet will undergo periodic dimming (positions 2 and 3) at intervals corresponding to the orbital period of the planet. Outside of the transit times (position 1) the star's brightness remains constant. The latitude of transit is given by the angle δ, with a perfect central transit corresponding to $\delta = 0$

Although the first exoplanets to be discovered were found through the Doppler technique, additional detection methods exist. Some discoveries have been made by direct imaging techniques, using a small, pinhead-sized occultation disk to block out the light from the parent star to reveal the faint reflected light of the planet (recall Fig. 2.11). Other new worlds have been discovered through gravitational lensing, where the planet induces a distinctive variation in the brightness of a star (as seen from Earth). Yet more, indeed, many more planets have been discovered by the transit method, whereby a planet moving in front of its parent star (in the observer's line of sight) causes distinct and periodic decreases in the stars brightness (Fig. 2.16). This method has produced dramatic results in recent years due to the Convection, Rotation and Transits (CoRoT) and Kepler spacecraft[47] missions conducted by the Centre National d'études Spatiales and NASA, respectively.

[47] Originally developed as the FRESIP (FRequency of Earth-Sized Inner Planets) mission the spacecraft was eventually named Kepler after Johannes Kepler (1571–1630), who not only discovered the basic laws of planetary motion but also pioneered the theory behind the design of modern-day optical telescopes.

If the planet within a transiting system has an orbital period P, a radius R_P and orbital radius a, then the transit time T to cross a star of radius R_S is

$$T = \frac{P}{\pi}\left(\frac{R_S \cos\delta + R_P}{a}\right) \tag{2.4}$$

For an alien observer monitoring the Sun when the transit of latitude is $\delta = 0$, the transit time will be of order 13 h for Earth, but just 5 h for Jupiter. These results bring out one of the advantages of the transit detection method. Although it is true that Jupiter is 11 times larger than Earth, its much greater distance from the Sun results in a much shorter transit time (by a factor of 2.6). For our transit monitoring alien observer, therefore, it is more likely that they will find Earth, which undergoes a 13-h transit once every year, than Jupiter, which undergoes a 5-h transit once ever 11.86 years. In general, the probability of observing a transit for randomly orientated systems is $P_{transit} = (R_S + R_P)/a$, when the longitude of transit is $\delta = 0$. In general, therefore, the probability that some alien observer somewhere within the galaxy might see Earth in transit across the Sun is $P_{transit} \approx 0.47$ %. The probability that Jupiter might be detected is nearly 5 times smaller, being $P_{transit} \approx 0.1$ %. Indeed, Venus has the highest probability of detection, by a random galactic observer, of all the planets within the Solar System, with $P_{transit} \approx 0.65$ %.

If we take the brightness (that is, measured flux f) of a star to be directly related to its cross-section surface area, then the flux ratio outside f_{out} and during a planet transit, f_{in} can be expressed as

$$\left(\frac{f_{out}}{f_{in}}\right) = \frac{R_S^2}{\left(R_S^2 - R_P^2\right)} \tag{2.5}$$

where R_S and R_P are the radii of the star and planet, respectively. Casting this in terms of a magnitude variation Δm (see Appendix 1 in this book), we have

$$\Delta m = m_{in} - m_{out} = -2.5\log\left[1 - \left(\frac{R_P}{R_S}\right)^2\right] \tag{2.6}$$

With respect to transit detection we see from equation (2.6), as would be expected, the larger the planet is compared to its parent star, the larger will the magnitude variation during a transit be (for a given orbital configuration). In the Solar System Jupiter is about a tenth the size of the Sun, and accordingly for a distant observer recording a transit $\Delta m = -0.01$; for the Earth, which is about a 1/100th the size of the Sun, $\Delta m = -0.0001$.

Although such flux (magnitude) variations are small, they are well within the domain of measurements with present-day technology, and this has allowed for the discovery of literally hundreds of new, small, Earth-sized planets. These planets, many hundreds of times less massive than Jupiter, are invisible to those surveys employing the Doppler technique, since their resultant reflex effect upon the parent star is too small to measure with current techniques.

In spite of a statistics-based failed prediction that Earth Mark II, literally an Earth-mass planet located 1 au away from a Sun-like star, would be discovered in May of 2011, it is no doubt just a matter of time before numerous Earth-mass planets situated 1 au from their parent Sun-like stars are discovered. This discovery, of course, will open up all manner of exciting opportunities to investigate the development of planetary atmospheres and possibly the evolution of life elsewhere in the galaxy.

In terms of possibly detecting planetary transits within the α Cen AB binary, the transit probabilities for us will be similar to those for an alien observer detecting Earth in orbit around the Sun. Formally, using Table 2.2 as our guide, the probabilities for detecting an Earth-sized planet having an orbital radius of 1 au are 0.2 % for α Cen A and 0.4 % for α Cen B. For Proxima the probability that an Earth-sized planet at 1 au will show transits is 0.07 %; an Earth-sized planet located in Proxima's habitability zone (to be discussed later below) with $a = 0.02$ au has a relatively high chance of showing transits, with $P_{transit} \approx 3.6$ %. The probability that α Cen Bb (described in more detail below) might show transits is not unreasonably low, at about 10 %.

Earth Mark II, as of this writing, still awaits discovery, but multiple planetary systems have already been found. The star υ Andromedae was the first such system to be discovered, and it sports four Jupiter-mass planets with orbital radii of 0.06, 0.83, 2.53 and 5.25 au. The star HD 69830 has three Neptune-mass planets. The star 55 Cancri has five planets (and an outer Kuiper Belt dust disc); and the star Kepler-11 has 6 Earth-mass planets in attendance – with orbits all squeezed into a region having an outer radius of 0.5 au. Compared to our Solar System five of the Kepler-11 planets have orbits smaller than that of Mercury. The range and variety of planetary systems is growing all the time, and the structure of our Solar System is beginning to look more and more routine, rather than exceptional, and this brings us back to consider, one last time, the possible usefulness of the Titius-Bode law.

As suggested earlier the power of the Titius-Bode law lies not in the fact that it explains any fundamental physical process but rather that stable planetary systems must satisfy certain conditions with respect to the orbits, spacing and resonances that exist between its members. A modern-day equivalent statement of the Titius-Bode law has been articulated by Rory Barnes and Richard Greenberg, both researchers at the University of Arizona. The Barnes and Greenberg statement addresses the dynamical nature of planet formation and planetary system stability, and argues that planetary systems tend to form in such a way that they are dynamically packed. This packed planetary system (PPS) hypothesis[48] essentially argues that if a planet *can* form at some specific location within the circumstellar disk about a newly forming star, then it *will* form.

Figure 2.17 shows the planetary spacing sequence for the star HD 10180, a Sun-like star located 39 pc away. For this particular

[48] This concept is incorporated into what has become known as the packed planetary system (PPS) hypothesis. This idea was first discussed in the research paper by R. Barnes and T. Quinn, "The (in)stability of Planetary Systems" (*Astrophysical Journal*, **611**, 494, 2004). Subsequent studies appear to have confirmed its veracity. It would indeed seem that if there are no specific physical reasons to stop a planet from forming in a stable region (i.e., gravitational resonances, gravitational migration and/or gravitational scattering), then a planet will form. Perhaps the ultimate application of the PPS is that by Sean Raymond (Bordeaux Observatory, France), who has constructed a "fantasy star system" composed of two red dwarf stars. By careful construction, Raymond is able to show that 60 Earth-mass planets might conceivably be situated, on dynamically stable orbits, within the systems two habitability zones. Various mathematical "tricks" were used to establish this number of habitable worlds, and though no non-physical principles were adopted, the probability of such a system forming naturally is essentially zero. The detection of any such massively packed planetary system could probably be taken as a clear sign that the work of a Kardashev II or III civilization (see Note 55 in Sect. 2.3) had been found. Details of Raymond's methods are given on the website www.obs.u-bordeaux1.fr/e3arths/raymond/.

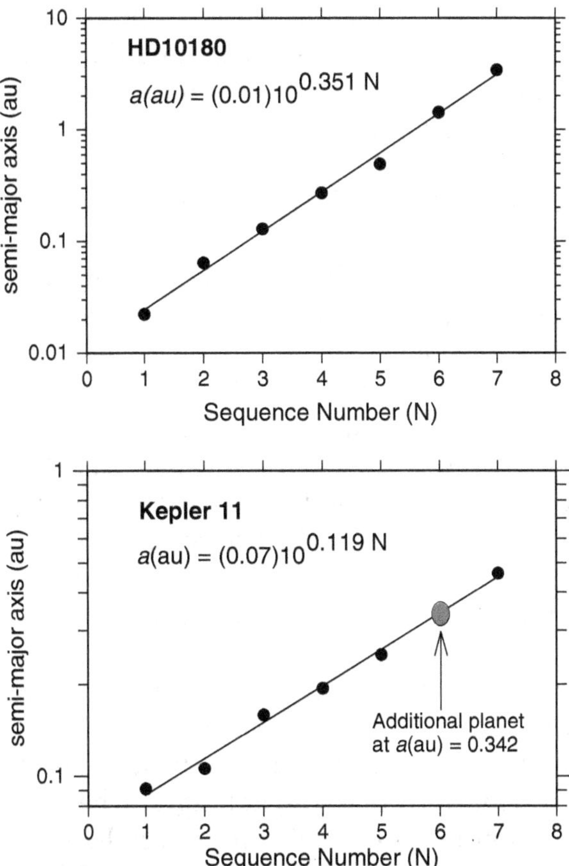

FIG. 2.17 The Titius-Bode-like laws for the stars HD 10180 and Kepler-11. The planetary system around HD 10180 appears to be a packed planetary system (PPS) – at least out to N = 7. For Kepler-11, however, if the PPS hypothesis genuinely holds true, then an additional planet should be located at sequence number N = 6

star the planet sequence appears to be complete for N = 1 to 7. There are no missed planets, and the system is fully packed. If more planets do exist in orbit around HD 10180, then they must have orbits larger than 6.4 au (corresponding to sequence numbers of 8 and above).

In contrast to HD 10180, the planetary spacings observed for Kepler-11 (also see Fig. 2.17) suggest that a planet is *missing* at N = 6. Under the PSS hypothesis this result suggests that the planet is not actually missing but rather not yet detected within the available dataset. Although the mass of the N = 6 planet in the Kepler-11 system cannot be predicted, other than it must be a terrestrial, low-mass planet, its orbital radius and period should be 0.342 au and 73 days respectively.

As we shall see in the next section, one planet has already been detected in orbit around α Cen B. Unfortunately, the manner in which the Titius-Bode law and/or the PPS hypothesis work requires the detection of at least three, and preferentially four or more, planets before any predictions about additional members can be made. We are currently at the impotent numerical end of the Titius-Bode sequence for α Cen B. If, and it is a very big if, it is assumed that the Titius-Bode law for α Cen B is similar to that for our Solar System and of the form $a(\mathrm{au}) = \eta \times \rho^N$, N = 0, 1, 2, ..., with $\eta = 0.02$ (making α Cen Bb the N = 1 planet in the sequence), then with $\rho = 2$ (approximately that derived for our Solar System) some five more planets (up to N = 6) might yet be squeezed into orbit around α Cen B. For N = 6, the orbital radius is about 1.28 au (corresponding to an orbital period of about 541 days); for N = 7, the orbital radius is 2.56 au, but this latter radius is beyond the stability limit for the star (as discussed earlier).

The numbers just presented are really pure fantasy and should not be taken seriously. They hint, at best, at what might be found. Given enough time and observational success, however, it is highly likely that some specific form of Titius-Bode-like laws will be derived for α Cen A, α Cen B and, quite possibly, Proxima Centauri.

2.12 Planets in the Divide

Within any binary system there are three zones where stable planetary orbits can exist: around each of the individual stars and around the binary system itself. Each set of configurations has been studied in the case of α Cen AB, and, for example, detailed numerical calculations conducted by Paul Wiegert and Matt Homan (then both located at the University of Toronto in Canada) in the late 1990s revealed that both stars can support stable planetary orbits out to about 4 au. Beyond this limit the gravitational perturbations of the non-parent star become significant, and a planet's orbit is rapidly destabilized.

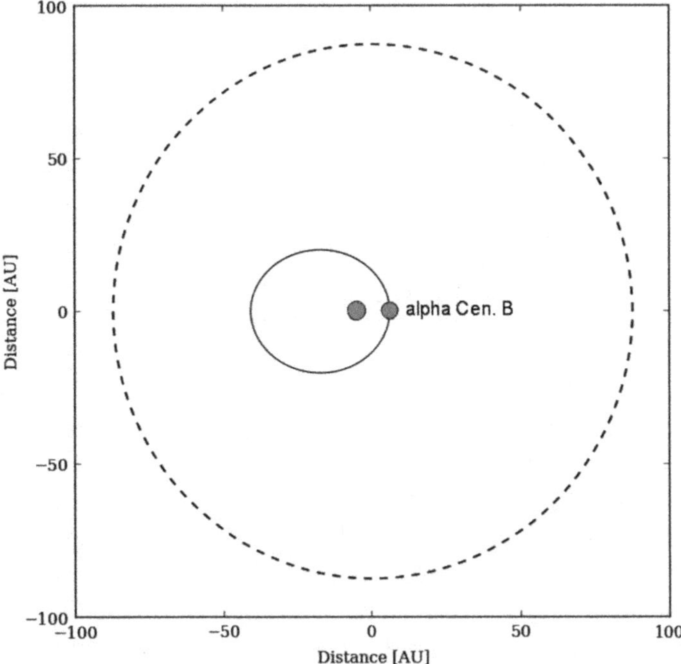

FIG. 2.18 Stability zones for co-planar planets in the α Cen AB system. The central ellipse indicates the orbit of α Cen B centered on α Cen A, the two small circular disks indicate the stability zones around each star ($a < 4$ au), while the large circle (*dashed line*) indicates the inner boundary of the outer orbital stability zone ($a > 80$ au) (see the web page calculator at: http://www.astro.twam.info/hz/)

Additionally, Wiegert and Homan showed that the inclination of the planetary orbits to that of the orbital plane of α Cen AB itself is very important. The 4 au stability limit applies if the plane of the planetary orbits is the same as that of two stars. As the orbital inclination increases, it turns out that the stability limit shrinks, and for 90° inclination orbits, the stability zone around each star is just 0.23 au in extent. Planetary orbits are also stable for distances further than 80 au from the barycenter of α Cen AB. Thomas Mueller and Nader Haghighpour (University of Tubingen, Germany) have recently developed[49] a web-based resource page that determines the stability and habitability zones for any specified binary system, and the result for α Cen AB are shown in Fig. 2.18.

[49] See the web page calculator at: http://www.astro.twam.info/hz/.

A growing number of exoplanets are being discovered in binary systems, with the planets either in orbit around one of the stellar components, as in the case of 55 Cancri A, or in orbit around both stars, as in the case of RR Caeli and Kepler-16ABb. In the case of the solar mass component 55 Cnc A, a total of five planets have been discovered with orbital radii in the range of 0.016–5.74 au.

As a point of interest (and later discussion) the half-Saturn-mass planet 55 Cnc Af was the first planet, outside of our own Solar System, to be found within the habitability zone of another star. The system RR Caeli is composed of an M spectral type red dwarf star and an evolved white dwarf. While these two "parent" stars are separated by just 0.008 au, Shengang Qian (Yunnan Observatory, China) and co-workers showed in a recent 2012 publication that they are both orbited by a four-times Jupiter-mass planet located at a distance of some 5.3 au. In contrast to the relatively expansive RR Caeli, the first planet to be discovered in a circumbinary orbit, Kepler-16ABb is much more compact. In this latter case a Saturn-mass planet orbits the central star system at a distance of 0.7 au. The K and M spectral-type stars that constitute the system's nucleus orbit each other at a distance slightly over 0.2 au. Discovered by Laurence Doyle (SETI Institute, at Mountain View, California) and co-workers in 2011, the entire Kepler-16ABb system could fit into the orbit of Venus within our Solar System.[50] There is no currently known binary system that has a set of attendant planets in orbit around each component, but such systems should exist, and it is really only a matter of time before the first one is going to be found.

The exoplanet surveys to date reveal that planetary systems can take on many different forms, and the imperative of nature appears to be that if a stable orbital region exists then a planet will be found within it. For the α Centauri system the lesson we learn from the exoplanet surveys is that there are potentially four regions in which planets might be found. Planets may exist in orbit around each of the stars, and in orbit between α Cen AB and Proxima.

---

[50] When first introduced the potential view from Kepler-16ABb was likened to that from the (imagined) planet Tatooine – the famous *Star Wars* (20th Century Fox, 1977) movie home of Luke Skywalker and the infamous womp rats.

2.13 First Look

"Look and yea shall find." Speculation and theoretical discussion are all well and good, but ultimately surveys and searches have to be made. What, after all is said and done, is the "ground truth"? As we saw earlier, the possibility that a planet, or at least a perturbing object, might exist around α Cen AB was invoked by Captain William S. Jacob in 1856. Although Jacob's speculation was in reality based upon uncertain data, it at least introduced the idea that otherwise invisible planets might be detectable around stars other than the Sun.

The first detailed search for possible planets in orbit around α Cen A and B was initiated by Michael Endl (University of Texas at Austin) and co-workers in 1992. Summarizing their radial velocity measurements some eight years later, Endl et al. found no planets, but they were able to place constraints upon the regions where planets might reside. Effectively, there can be no planets more massive than 1 $M_{Jupiter}$ within 2 au of α Cen A, and no planets more massive than 2 $M_{Jupiter}$ within 4 au. For α Cen B, there are no planets more massive than 1.5 $M_{Jupiter}$ within 2 au, and no planets more massive than 2.5 $M_{Jupiter}$ within 4 au.

In terms of circumbinary planets, a deep image survey in the region immediately surrounding α Cen AB by Pierre Kervella (ESO, Garching) and Frederic Thevenin (Observatoire de la Côte d'Azur, France) revealed no co-moving objects more massive than 15 $M_{Jupiter}$ out to distances of order 100–300 au. These initial survey results effectively indicate that no large, multiple Jupiter-mass, planets exist in orbit around either α Cen A or B. Such planets may yet be discovered, however, as circumbinary, cold, Jupiter objects. If there are multiple planets in orbit about α Cen A and/or B then it is to be expected that they will be sub-Jupiter in mass, and this accordingly sets the requisite resolution limit for future radial velocity surveys to be better than at least 1 m/s (but also see below).

Proxima has also been deep searched for possible planetary companions. Early observations date back to at least 1981, when R. F. Jameson (University of Leicester, England) and co-workers used the 3.8-m United Kingdom Infrared Telescope (UKIRT) on Mauna Kea to search for brown dwarf companions to nearby stars.

Proxima was on their survey list, but no hint of any large companion was found. The possibility of Proxima having a brown dwarf companion was raised again in 1998. At this time, Al Schultz (STSI, Baltimore) and co-workers reported that they had obtained observations with the Hubble Space Telescope's faint object spectrograph that hinted at a possible brown dwarf companion some 0.5 au from Proxima. This tentative detection, however, did not survive for very long, and within in a year new data showed that no such companion existed. Indeed, using astrometric data provided by the Hubble Space Telescope's guidance system, Fritz Benedict and co-workers were able to show in a 1999 publication that no planet more massive than 0.8 $M_{Jupiter}$ with an orbital period of between 1 and 1,000 days is in existence around Proxima.

Martin Küster (Max Planck Institut für Astronomie, Heidelberg) and co-workers have additionally shown, in another 1999 publication, that no planets within the mass range of 1.1–22 $M_{Jupiter}$ with orbital periods of between 0.75 and 3,000 days could exist in orbit around Proxima. Furthermore, Endl and Küster were able to show, in a 2008 publication, that no planet having a mass greater than about twice that of Earth can exist within the habitability zone (the region between 0.02 and 0.05 au) of Proxima. Figure 2.19 provides a graphical summary of the survey results applied to date.

Figure 2.19 indicates that at the present time there is still a large parameter space yet to be explored when it comes to detecting possible planets within the α Centauri system. We can be reasonably sure that no Jupiter-mass planets exist in orbit around either α Cen A or B. Such objects may yet exist in orbit around Proxima, but they must have orbital radii greater than 0.01 au. For planet masses less than that of Jupiter, however, the entire orbital stability zones of α Cen A and B have yet to be studied. At least one Earth-mass planet exists in orbit around α Cen B, and there is no specific reason to rule out the possible existence of others. At the present time it is possible that Earth-mass planets exist, and await discovery, within the habitability zones surrounding each of the three stars in the α Centaurus system.

With respect to present-day technology it is not possible to directly measure the reflex velocity induced by an Earth-mass planet situated at 1 au away from either α Cen A or B. Indeed, to achieve the latter at least an order of magnitude improvement will

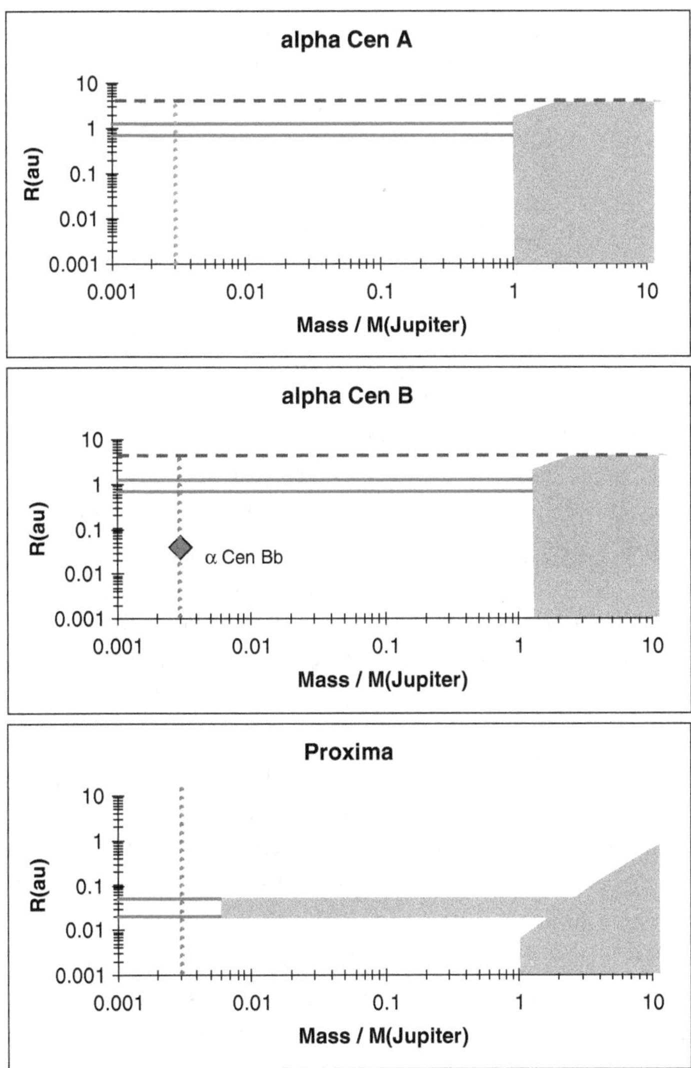

FIG. 2.19 Planetary mass and orbital radius limits for the α Centauri system. In the *top* and *middle* diagrams the *horizontal dashed line* at 4 au indicates the stability limit for planetary orbits. In all three diagrams the *solid horizontal lines* indicate the extent of the habitability zone. The *shaded regions* indicate the presently excluded zones for planets of a given mass and location. The *vertical dotted line* indicates the location at which Earth-mass planets could be found

be required in the Doppler velocity measurement techniques (see later). Improving the radial velocity precision is only part of the story, however. As we saw earlier both α Cen A and B are active, Sun-like stars, and this activity will induce additional "noise" into the radial velocity data. Moving forwards, therefore, is not only a matter of improving the precision with which radial velocity measurements are made, but also about knowing the properties and behaviors of the parent stars.

2.14 The Signal in the Noise

In order to find Earth-mass planets in orbit around either α Cen A or B, the mantras of "pile the data high" and "pile the data deep" might meaningfully be applied. Indeed, to find a planet in orbit around either one of these stars is to literally hunt for the proverbial needle in a haystack. The radial velocity data is going to be noisy, and it will contain multiple sources of variation – some periodic and some not. Before, therefore, the reflex motion of the star, due to the presence of a planetary companion, might be evident all the additional source terms will need to be extracted out from the observational dataset. The idea here is that by subtracting out the known variations, what is left, with luck, will be the signal of a planet.

To deep search α Centauri for planets was always going to be a Herculean task, but it was by painstakingly sifting out the noise that a team of 11 researchers, under the lead authorship of then graduate student Xavier Dumusque (Observatoire de Genève, Switzerland), were able to announce the discovery of α Cen Bb in late 2012.[51] To begin with, the team of observers gathered radial velocity data on α Cen B over a 3-year time interval – between February 2008 and July 2011 – and a total 459 radial velocity measurements were obtained with the HARPS spectrograph (recall Fig. 2.15).

It was decided from the outset of this study to concentrate on α Cen B, rather than α Cen A, since the former is slightly less massive and accordingly the radial velocity variation due to a terrestrial planet (for a given orbit; see Eq. 2.3) will be stronger, and this is important since the expected variations are going to be small – indeed, less than a meter per second. Given the known sources

[51] The results were announced in the prestigious journal *Nature* on November 8, 2012. Perhaps surprisingly, no byline or information was given on the journal's front cover about the remarkable discovery paper that was contained inside.

that can induce Doppler variations, the measured radial velocity (RV) can be thought of as the summation of at least eight terms. Accordingly,

$$\begin{aligned} RV = & RV(\text{binary}) + RV(\text{rotation}) + RV(\text{magnetic cycle}) + RV(\text{oscillations}) \\ & + RV(\text{granulation}) + RV(\text{instrument noise}) + RV(\text{Earth}) + RV(\text{planet}) \end{aligned}$$

where the terms in brackets indicate the mechanism responsible for the radial velocity variation.

The trick and also the time-consuming part of the analysis, once having gathered the RV data, is to subtract out the unwanted RV terms. Fortunately most of the "noise" terms in the RV data are reasonably well understood, and they can accordingly be removed in a (reasonably) reliable fashion. The binary period of α Cen AB, for example, is certainly well known (79.91 years), and its contribution over the 3-year observational cycle is easily removed. Likewise, the rotation period (38.7 days) of α Cen B is also well determined, and its modulation effects can be readily subtracted out as well. The magnetic cycle for α Cen B, over which time the surface dark spot number will vary, is estimated to be of order 9 years, and to subtract out this term Dumusque and co-workers collected chromospheric activity data at the same time as the RV measurements were made.

With the chromospheric activity index data in place a radial velocity correction for the magnetic cycle activity could be made. This is perhaps the least well understood part of the process. The corrections due to surface oscillations (typically having periods of about four minutes) and surface granulation (due to rising and falling surface convection cells) were averaged out by choosing a data-collecting exposure time of ten minutes. The idea here is that over such exposure times the surface oscillations and granulation effects should average out to a very small effect. The instrumental noise is subtracted out by carefully measuring the precision and functioning of the HARPS spectrograph over time, and the final correction RV(Earth) is related to the reflex motion of the Sun (recall Fig. 2.13) and the subsequent small velocity variation that it causes in Earth's velocity towards α Cen B.

Examining each of the various RV source terms in detail, Dumusque and co-workers developed a 23-free parameter, time variable correction term to subtract out from the measured RV data. Sieving, then, the corrected radial velocity dataset (Fig. 2.20)

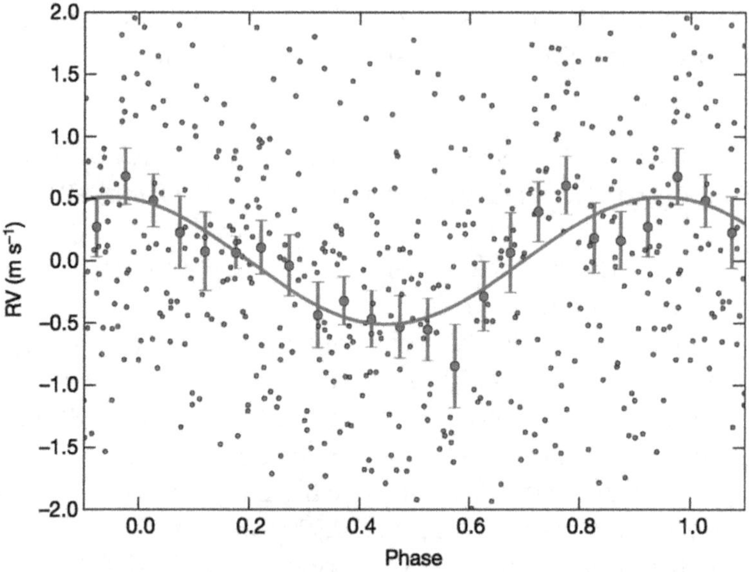

FIG. 2.20 The corrected radial velocity data (*green dots*) for α Cen B. The data points have been folded according to the period of the planet α Cen Bb (3.2357 days) and the *red dots* show time-averaged means. The best fit to the radial velocity data curve is shown in *red* (Image from doi:10.1038/nature11572. Used with permission)

for periodic behavior they found two possible signals at 3.2357 and 0.762 days. Continued statistical testing of the data eventually indicated that only the 3.2357 day signal was real, with the probability that this is a false positive result (meaning it is entirely due to noise within the data) being estimated at about 1 in 500.

The induced reflex velocity variation due to α Cen Bb has an amplitude of just 0.51 m per second (indicated by the red curve in Fig. 2.20), and its (minimum) mass and orbital radius are 1.13 Earth masses and 0.04 au, respectively. The analysis by Dumusque and co-workers provides a minimum mass for α Cen Bb, since it is not yet known what the orbital inclination of the system is. The effect of varying the inclination of the orbit with respect to our line of sight is illustrated in Fig. 2.21. The published value of 1.13 Earth-masses for α Cen Bb is based on the assumption that we are seeing right into the orbit (corresponding to an inclination of exactly 90°). If the inclination is exactly 90° then transits will also take place in our line of sight, but no such signal has as yet been recorded.

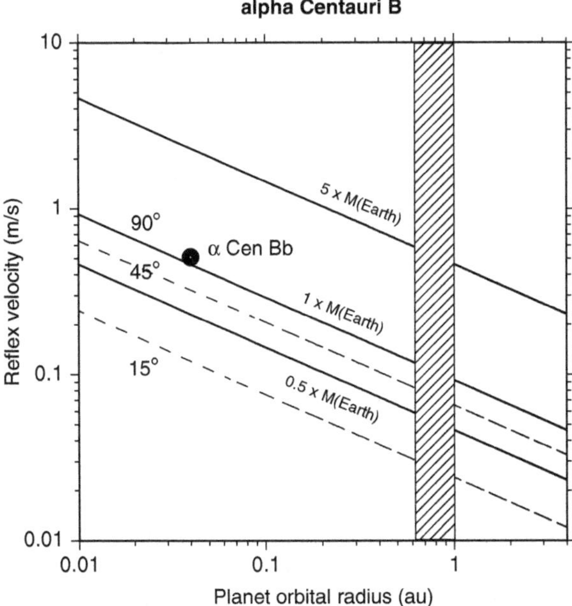

FIG. 2.21 Reflex velocity versus orbital radius for various mass planets. The solid lines correspond to orbital inclinations of 90°. The two *dashed lines* labeled 45° and 15° correspond to a 1 Earth-mass planet and illustrate the effect of reducing the orbital inclination

For orbital inclinations smaller than 90°, the deduced mass for α Cen Bb will increase. While α Cen Bb appears to be an Earth-analog planet it is not an Earth Mark II since it orbits α Cen B well inside of its habitability zone (Fig. 2.22 and see below) (shaded region in Fig. 2.22 and see below).

A few words of caution are now due. The subtraction procedure developed by Dumusque and co-workers, although performed to the highest standards of analysis and rigor, may nonetheless contain some subtle effect that resulted in the apparent periodic planet signal at 3.2357 days. Reproducibility of results being one of the most important cornerstones of science behooves us therefore to record that the detection of α Cen Bb is still preliminary, and affirmation of its true existence awaits efforts by other research groups using independent datasets. Not only this, additional analysis of the subtraction procedure itself is required; the scheme used by Dumusque et al. is not necessarily unique and/or the best to apply. Indeed, Artie Hatzes (Thuringian State Observatory,

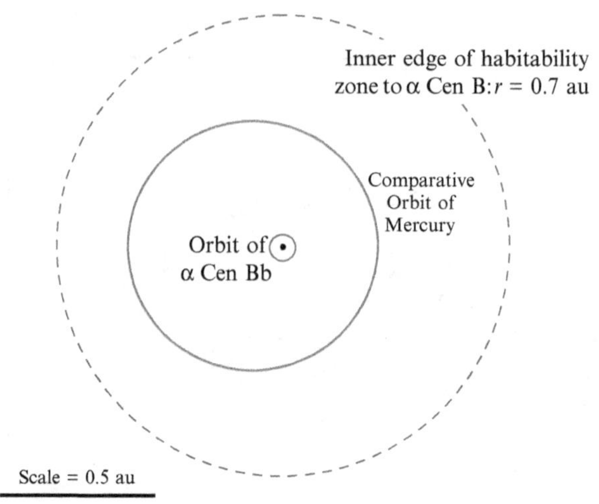

FIG. 2.22 The orbit of α Cen Bb shown in comparison to our Solar System. The 0.04 au orbital radius of α Cen Bb places it some ten times closer to α Cen B than Mercury is to our Sun. The location of the inner boundary ($r = 0.7$ au) of the habitability zone around α Cen B is shown by the *dashed circle*. The scale bar indicates a distance of 0.5 au

Germany) has re-analyzed the α Cen B RV dataset obtained by Dumusque and co-workers and finds that different results might occur if the raw data is divided up and analyzed in slightly different, but perfectly allowable, ways. "The detected *planet* seemed to be highly sensitive to the details in how the activity variations are removed," concluded Hatzes in a May 21, 2013, preprint (arxiv.org/pdf/1305.4960v1.pdf). Hatzes is not claiming in his analysis that Dumusque and co-workers are wrong, or that α Cen Bb does not exist. What he is very appropriately saying is that more, indeed, much more observational data is required to fully bring out the all-important planet signal.

The observational technique used by Dumusque and co-workers is that of Doppler shift measurements. Such observations are being used to reveal the reflex motion of α Cen B. An alternative, perhaps more direct, approach to looking for (confirming) α Cen Bb is to search for transit variations. In this case, as discussed earlier, slight drops in the brightness of α Cen B would be evident each time (once per orbit) any planet moved across its disk in our line of sight. The key point, however, is whether the orbit of any planet is orientated such that transits might be observed from Earth. It was noted earlier that the probability that a randomly orientated system might show planetary transits is $P_{transit} = (R_S + R_P)/a$. Adopting the appropriate values for an Earth-sized planet in orbit around α Cen B with an orbital radius of 0.04 au, we have $P_{transit} \sim 10$ %.

The possibility of observing planetary transits is not high for α Cen B, but then neither is it zero. Accordingly, an international team of observers, including Xavier Dumusque, has used the Hubble Space Telescope to monitor α Cen B for the subtle brightness dips due to a planetary companion in transit. Lead investigator for the HST observations is David Ehrenreich (Observatoire de Genève, Switzerland) and the data-gathering run was conducted during a 26-h observing window (corresponding to sixteen orbits of the spacecraft) in July of 2013. The precision of the observations is such that the light dips due to an Earth-sized planet in transit across α Cen B should be detectable (if the orbital plane is favorable for observing transits from Earth). As of this writing no announcements have been made concerning the results of the HST study, but if a transit is captured then not only will this confirm the existence of α Cen Bb, it will also provide a direct measure of the planet's radius. Once the radius is known then the density of the planet can be determined, and this will provide important information about the exact composition and structure of α Cen Bb. Additionally, the HST observations will potentially provide a spectrum of α Cen Bb, and although it likely has no extensive atmosphere (see below) it may leave a vapor trail (possibly of surface-evolved sodium) behind it as it orbits α Cen B.

What would it be like to stand on the surface of α Cen Bb? The answer to this question partly depends upon which hemisphere you might be located, but either way the prospects would be decidedly grim. "Farewell happy fields where joy forever dwells: Hail horrors, hail infernal world, and thou profoundest Hell" – so writes John Milton (1608–1674) in *Paradise Lost*, and in many ways this sums up the outlook for an observer on α Cen Bb. As we shall see below, α Cen Bb is sufficiently close to α Cen B that it will be tidally locked. This dictates that one hemisphere will experience perpetual daylight, always facing towards α Cen Bb, while the other hemisphere will be cast in permanent night. The one hemisphere will be hellishly hot, while the other will be hellishly cold.

On the daylight hemisphere α Cen B will loom large with an angular diameter of 11.5° on the sky. This is some 23 times larger than the Sun appears to us from Earth. The influx of surface energy from α Cen Bb will be a staggering 430,000 W per m^2 (312 times larger than the Sun's energy flux at the top of Earth's atmosphere), and the resultant surface temperature will be around 1,200 K. The very rocks on the daylight hemisphere of α Cen Bb will ooze and fold, and no solid land masses will exist to support a would-be surface observer.

Moving towards the permanent nighttime hemisphere α Cen B will fall lower and lower in the sky, eventually disappearing, never thereafter to rise. From here on in the temperature will drop precipitously to the biting cold of interstellar space. No sunrises to heat the ground and no atmosphere exists to circulate any of the dayside heat. From the permanent night hemisphere, the brightest object in the sky will be α Cen A – which at intervals of 80 years (see Appendix 3 in this book) will rise to a maximum brightness of magnitude −22 – making it technically just a little fainter than the Sun appears to us on Earth. Its angular diameter, however, will be about a tenth that of the Sun (as seen by us on Earth). Indeed, from α Cen Bb, α Cen A will shine like a piercing diamond in the sky. For all of the pointillist brilliance of α Cen A, however, the nighttime hemisphere of α Cen Bb will be wrapped within the frozen embrace of a withering cold. These are not the Elysian fields. In short, α Cen Bb is not a place you would ever want to visit – other than, that is, by using a means of virtual telepresence.

2.15 Bend It Like Proxima

In addition to employing Doppler and transit survey methods to find exoplanetary systems, a third search method, based upon the gravitational bending of light rays, has also been successfully developed to find new worlds. The monitoring requirements for the gravitational microlensing technique are essentially the exact opposite of those used in the transit method. Rather than looking for a periodic decrease in the brightness of a star, it is a characteristic brightness increase that is looked for.

Not only does the brightness of a background source star (or galaxy) increase during a lensing event, the apparent position of

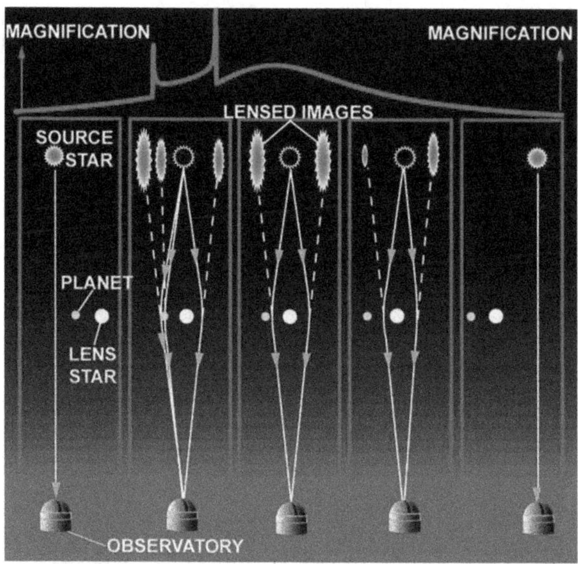

FIG. 2.23 Schematic diagram for the gravitational bending of light by a "lens" star situated between a "source" star (or galaxy) and the observer. If a planet is in orbit around the lensing star then it can produce an additional microlensing effect (Image by Dave Bennett, University of Notre Dame. Used with permission)

the lensed star (or galaxy) also shifts slightly in the sky. The amount of shift depends upon the mass of the lensing object; the bigger the mass, the greater the positional offset produced. The essential geometry behind the gravitational lensing technique is illustrated in Fig. 2.23.

The gravitational microlensing technique of planetary detection typically comes into its own when very distant star fields are being monitored, since such fields provide a large number of potential sources to lens. As of the time of this writing, with 1,047 known exoplanets having been detected in 794 planetary systems, some 25 exoplanets in 23 planetary systems have been discovered through the microlensing effect.

The first exoplanet detected by the lensing technique has the ungainly catalog name of[52] OGLE-2003-BLG-235/MOA-2003-BLG-53b. This planet has a mass some 2.6 times greater than that

[52] In the ever-more chaotic and contrived world of acronyms, we have Optical Gravitational Lensing Experiment (OGLE) and Microlensing Observations in Astrophysics (MOA).

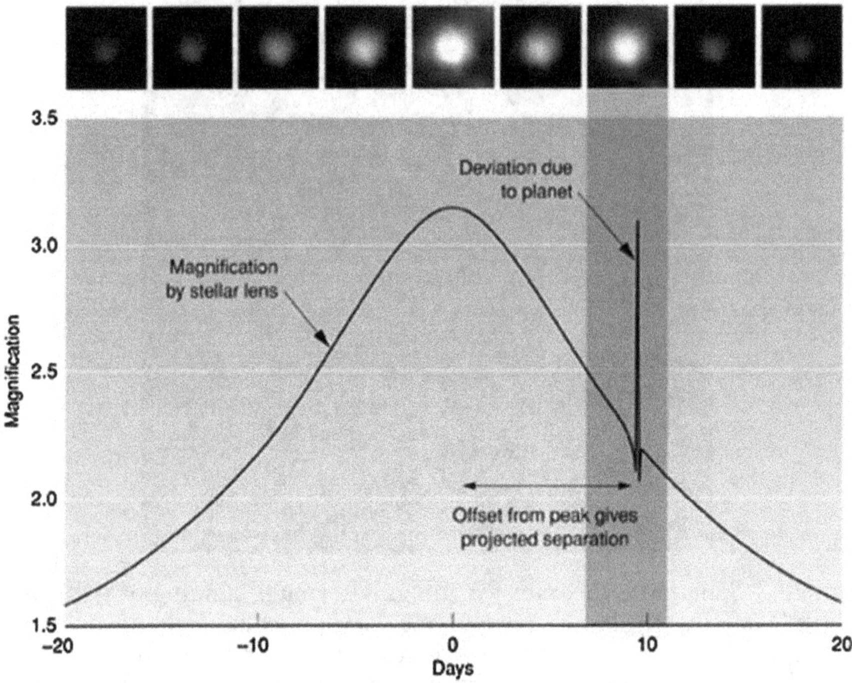

FIG. 2.24 Schematic variation in light curve brightness as the lens star moves in front of a source star during a gravitational microlensing event. The short spike in magnification is due to the presence of a planet in orbit around the lensing star (Image courtesy of the Lawrence Livermore National Laboratory)

of Jupiter, and it orbits a K spectral-type parent star (slightly less massive than the Sun) at a distance of some 4 au. Discovered in 2003, the planet and its parent star are situated at a distance of about 5,800 pc from the Sun in the galactic bulge surrounding the very center of our Milky Way Galaxy.

Key to the success of the microlensing technique (as illustrated in Fig. 2.23) is the alignment of a foreground object (the lens) with a distant light source provided by another star or a distant galaxy. For a very close alignment in the observer's line of sight the gravitational field of the lensing object will produce multiple and brighter images of the background source, and it is this brightening, typically lasting from weeks to months, that is searched for. If the foreground lensing object is a star with a planet, then this can, if the geometry is just right, result in a shortlived but even greater brightening of the background source (Fig. 2.24).

Just as with the transit surveys, the planet-induced brightening spike, however, will only last for a few hours, and this, of course, acts against the chances of it being detected unless near full-time monitoring techniques are employed.

Two important results to have appeared from the various microlensing surveys run to the present time are the identification of a large population of rogue planets within the Milky Way, and the deduction that, "Stars are orbited by planets as a rule, rather than the exception."[53] Both of these results are remarkable, and they present a dramatic shift in historical thinking. First, the latter of the two findings tells us that planet formation is both a natural part and a common outcome of the star formation process. Indeed, a 2012 microlensing study report published in the journal *Nature* by Cassan (Institut d'Astophysique de Paris) and co-workers indicates that on average every star in the Milky Way Galaxy has 1.6 ± 0.8 planets in the mass range from 5 Earth masses to 10 Jupiter masses, with orbital radii between 0.5 and 10 au. The result, that unbound Jupiter-mass planets not only exist but actually outnumber stars by a factor of approximately two to one, was also a microlensing survey result, this time published in the journal *Nature* by Takahiro Sumi (Osaka University) and MOA[54] co-workers in 2011. These rogue planets, no doubt, formed within the gas and dust disks that are associated with newly forming stars, but due to the combined processes of planet migration and gravitational scattering interactions have been launched onto lonely, unbound trajectories that carry them through the cold of interstellar space.

In contrast to the paths followed by rogue planets, Proxima has a well-known proper motion path across the sky, and accordingly systematic searches of star catalogs can be used to predict the exact times when lensing events, with Proxima being the lens, might occur. Samir Salim and Andrew Gold, at Ohio State University in Columbus, performed just such a search in late 1999 and found three occasions (in 2006, 2010 and 2013) on which Proxima

---

[53] This quotation is taken from the research paper by A. Cassan et al., "One or more bound planets per Milky Way star from microlensing observations" (*Nature*, **481**, 167, 2012).

[54] See, T. Sumi et al., "Unbound or distant planetary mass population detected by gravitational microlensing." – paper available at arxiv.org/abs/1105.4544v1. MOA (see Ref. 52) is a long-running and highly successful collaboration between astronomers in New Zealand and Japan using gravitational lensing techniques to study dark matter, exoplanets and stellar atmospheres.

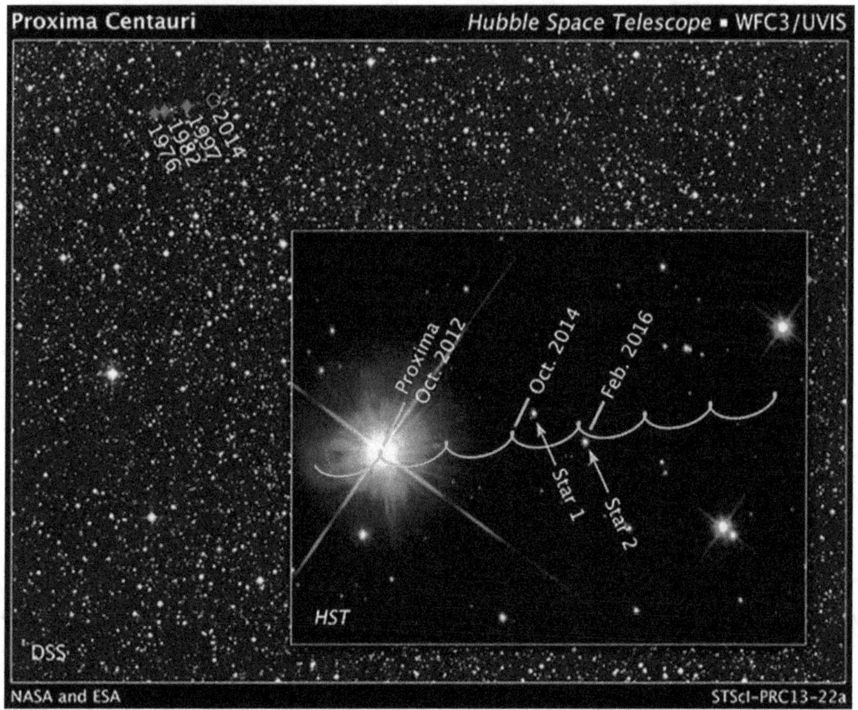

FIG. 2.25 The proper motion track (*lower right-hand corner*) of Proxima Centauri over the time span 2011–2018, showing the times and sky locations where the background star occultations will take place in 2014 and 2016. The background star field and the locations of Proxima (since 1976) are shown in the *upper left-hand corner* of the image (Image courtesy of ISTS and NASA)

would lens background stars. Unfortunately, none of these predicted events were monitored. A similar study led by Kailash Sahu, of the Space Telescope Science Institute in Baltimore, however, has additionally revealed that Proxima lensing events will take place in October 2014 and February 2016. These latter two events nicely bookend a remarkable set of centennial celebrations, with 2015 not only marking the 100th anniversary of Proxima's discovery by Robert Innis but also the 100th anniversary of Einstein's first public presentation on general relativity (the scientific theory behind gravitational lensing) at a meeting of the Prussian Academy of Sciences in Berlin.

The proper motion track of Proxima from 2014 to 2018 is shown in Fig. 2.25 (recall also Fig. 1.12 for α Cen A and B). Precise measurement of the shift, anticipated to be between 0.5 and 1.5 milliarcseconds, in the positions of the lensed stars, over total event time intervals of just a few hours, will enable the direct determination of Proxima's mass to unprecedented accuracy. Indeed, the gravitational lensing method is the only direct method available to astronomer by which a single star's mass can be determined. Normally a star must be located within a binary system (such as in α Cen AB) for its mass to be derived. In addition, if the geometry is just right, and if luck is with us, the lensing events could also reveal the presence of close-in, Earth-sized planetary companions to Proxima. Importantly, and in contrast to the normal situation in which it is purely random chance that determines whether a lensing event will take place, and when it does it is a one-off affair, with Proxima its known path in the sky means that multiple and predictable lensing events take place. And, while one particular lensing encounter might not have the correct geometry for the detection of close-in planets, another one just might.

2.16 The Sweet Spot

Assuming that multiple numbers of planets are eventually discovered within the α Centauri system, then one of the most intriguing follow-on questions that can be asked is, "Could life have evolved upon any one or more of them?". This is indeed a profound historical question, not just of α Centauri but of all stars within the Milky Way. Once again, however, we live in a remarkable epoch where observational studies can potentially provide us with a direct answer to the question. And, while as yet we have only a poor understanding about the origins of life, that is, the workings of the initial 'spark' that changes a collection of inanimate atoms and molecules into a self-regulating, reproducing, conscious living entity, we do know at least some of the conditions required for that primordial 'spark' to come about.

On Planet Earth life has been maintained and protected by the existence of an atmosphere and a liquid ocean, and the conditions for these two features to exist over billion-year timespans are well understood in terms of planet size, atmospheric constitution and distance from the Sun. It is through this basic understanding that the concept of the habitability zone (HZ) has come about.[55] For Sun-like stars the region in which a terrestrial mass planet might support an atmosphere capable of providing sufficient pressure for liquid water to exist at its surface has an inner boundary radius of about 0.95 au and an outer boundary radius of about 1.4 au. In approximate terms, the locations of the inner and outer habitability zone boundaries are proportional to the square root of the parent star's luminosity:

$$r_{inner}(au) = 0.95\sqrt{L} \text{ and } r_{outer}(au) = 1.4\sqrt{L} \tag{2.7}$$

where the luminosity is expressed in solar units.

As we have seen earlier α Cen A and B are approximate Sun analogs, so their habitability zones will be similar in extent to that for our Solar System (Fig. 2.26). Since α Cen A is more luminous than the Sun (Table 2.2), however, its habitability zone is displaced further outward than that for our Solar System, and indeed, the straight translation of Earth to α Cen A at 1 AU would place it in a region too hot for liquid water to exist on its surface. A translation of Earth to an orbit around α Cen B would place it close to the outer, cold edge of the habitability zone, making it a somewhat less hospitable place for life (as we know it) to thrive. Not only would Earth be habitable at 1 au from α Cen B, but so, too, would a transplanted Venus, since at 0.724 au it would sit beyond the innermost, hot edge of the habitability zone. A straight transplantation of Earth to an orbit around Proxima would place it well outside of the habitability zone, resulting in a frozen and lifeless world.

The line marked tidal lock radius in Fig. 2.26 corresponds to the boundary interior to which an Earth-sized planet would rotate in such a fashion that the same hemisphere of a planet will always face the parent star. The line, as shown, corresponds to the distance for tidal lock to have occurred in a time of six billion years. The time t_{lock} for the tidal locking condition to come about is dependent upon the planet's orbital radius a and the mass of the parent star M – approximately, for Earth-mass planets,

$$t_{lock}(\text{yrs}) = 10^{12}\left(a^6 / M^2\right) \tag{2.8}$$

---

[55] A web-based calculator for estimating the inner and outer habitability zone radii has been developed at the University of Washington. It can be accessed at http://depts.washington.edu/naivpl/content/hz-calculator.

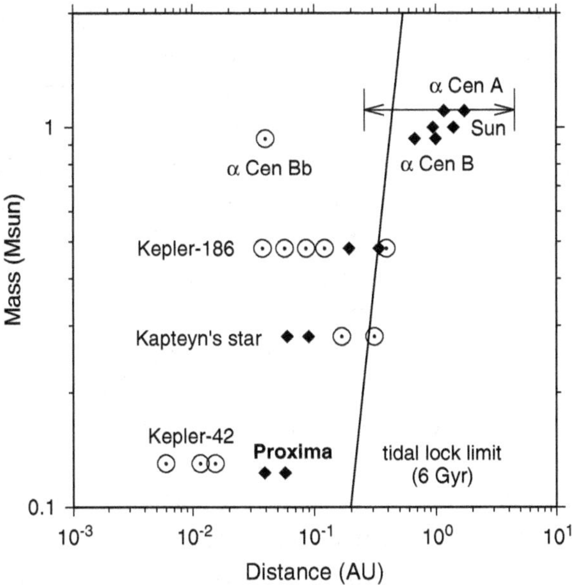

FIG. 2.26 The inner and outer edges of the habitability zone (*filled diamonds*) for the stars in the α Centauri system compared to that for the Sun and Solar System. The *diagonal line* indicates the boundary at which an Earth sized planet would become tidally locked after 6 Gyr – the age of the α Centauri system. The *dotted circles* indicate the locations of planets within the Solar System (Mercury to Jupiter), α Cen Bb and the Proxima analog system Kepler-42. The planet locations for Kepler 186 and Kapteyn's stars are also shown. The *horizontal line* through the α Cen A habitability zone points shows the outer stability radius for co-planar (4 au) and 90° (0.23 au) inclination planetary orbits

where a is expressed in astronomical units and M in solar masses.

Equation 2.8 indicates that the closer a planet is to its parent star, the shorter is the tidal locking time. Likewise, for a given orbital radius, the more massive the parent star, the shorter is the tidal locking time. Clearly, again from Fig. 2.26, α Cen Bb is situated well inside of the tidally locked region. Any Earth-sized planets that might be located within the habitability zones around α Cen A and/or B, however, will, on the other hand, not have reached a tidally locked state.

It is not fully clear yet what the consequences of tidal lock might be on a potentially habitable planet – i.e., one with an atmosphere. It is clear, however, that the heating of one planetary hemisphere and not the other will have a dramatic effect on atmospheric structure, wind circulation and surface temperature distribution. As to whether such planets can support life is still unclear. Where such affects will be critical is for any Earth-sized planets that might orbit Proxima Centauri. As Fig. 2.26 indicates, any planets that might be situated within the habitability zone around Proxima will be tidally locked, and consequently the possible existence of and indeed the very evolution of any associated biosphere can only be speculated on at the present time. The scientific community right now appears to be split as to the exact consequences of tidal locking on the habitability of a planet; some researchers argue that such conditions must of necessity preclude the existence of any surface habitability zones, while others suggest that regions situated close to the day-night divide boundary might just support an active biosphere.

Although the habitability zone will move outwards as Proxima ages and its luminosity increases, at no time will the outermost edge of the habitability zone move beyond the tidal lock radius. Planets may well exist within the canonical habitability zone around Proxima, but it is far from clear as to whether we should expect to ultimately see the evolution and/or presence of any indigenous life forms.

The announcement of the first Earth-sized planet to be discovered within the habitability zone of a red dwarf star, planet Kepler-186f, was made in April 2014. This planet, which is one of five detected in the system, is located towards the extreme outer edge of the habitability zone of Kepler-186 (see Fig. 2.26), and it receives about the same energy flux as Mars from our Sun. To stay warm enough for water to exist on Kepler-186f, therefore, it would need something like a dense CO_2 atmosphere to provide a strong greenhouse heating effect. Such an atmosphere might conceivably be produced through volcanic outgassing, and more importantly, may also be detectable with next generation instruments.

Although in principle low-mass, Earth-like planets can exist around Proxima in orbits with radii of many astronomical units, the same cannot be said for α Cen A and or B. As discussed earlier, the binary companionship of these two stars limits the size of the stability region to about 4 au (recall Fig. 2.18). This stability region, however, encompasses the habitability zone, and accordingly

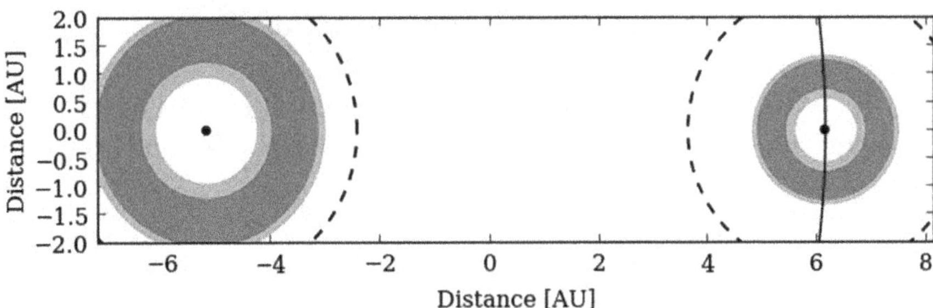

Fig. 2.27 The habitability zones (*shaded*) of α Cen A (*left*) and B (*right*). The scale is in astronomical units, and the *dashed curves* indicate the limit for stable planetary orbits. Although the extent of the habitability zones does not change during the orbit, the figure shows α Cen A and B at their closest approach (Image derived from Mueller and Haghighipour. (See the web page calculator at: http://www.astro.twam.info/hz/.) Used with permission)

Earth Mark II may yet exist within the closest star system to us. Figure 2.27 shows the expected extent of the habitability zones associated with α Cen A and B.

2.17 Alpha Centauri C?

From the very first moment of its discovery Proxima Centauri presented astronomers with a puzzle. First, Robert Innis noted in his initial communication that Proxima had a proper motion almost identical to that of α-Centauri, and he suggested that the two systems might be associated, making up thereby a small, common proper motion star cluster.

When Joan Voûte discovered that Proxima was at essentially the same distance as α Centauri he questioned, "Are they physically connected or members of the same drift?" Ninety years on from its first airing Voûte's question has still not been resolved. At issue, specifically, is the question, if Proxima is gravitationally bound to α Centauri, then what is its orbital path around α Cen AB? For Proxima to be moving in a bound (that is, elliptical and periodic) orbit around α Cen AB the total energy E of the system must be less than zero. If $E \geq 0$ then Proxima cannot be gravitationally bound to α Cen AB, and its relative closeness to α Cen AB must be a pure (and remarkable) coincidence of the present epoch. The total energy E of the system will be the sum of the kinetic energy and the gravitational potential energy as measured from the system's (α Cen AB + Proxima) center of mass. Accordingly, for a bound orbit it is required that:

$$E = \frac{1}{2}\left(\frac{M_{AB}M_{Prox}}{M_{AB}+M_{Prox}}\right)V^2 - G\frac{M_{AB}M_{Prox}}{r} < 0 \tag{2.9}$$

where M_{AB} is the combined mass of α Cen A and α Cen B, M_{Prox} is the mass of Proxima, V is the relative velocity of Proxima about α Cen AB, r is the relative distance of Proxima, and G is the gravitational constant.

With $M = M_{AB} + M_{Prox}$ Eq. 2.9 can be recast to set an upper limit on the relative velocity of Proxima:

$$V < \sqrt{\frac{2GM}{r}} \tag{2.10}$$

Since all of the quantities in Eq. 2.10 are measurable, the question now is what is actually observed. The result depends upon the observed masses of the stars, their separation r (which is based on their angular separation in the sky and the system parallax) and the relative velocity of Proxima compared to α Cen AB.

Dealing with the right-hand side of Eq. 2.10 first, the condition on the relative velocity based on the measured masses and separations is $V < 0.399 \pm 0.012$ km/s.[56] In contrast to this number, the measured velocity of Proxima relative to α Cen AB is only poorly known. Pourbaix and co-workers (see Appendix 3 in this book) have measured to high precision the radial velocity of α Cen AB and find $V_{AB} = -22.445 \pm 0.002$ km/s. At the present time, however, the best estimate for the radial velocity of Proxima is $V_{Prox} = -21.8 \pm 0.2$ km/s, and accordingly $V = V_{AB} - V_{Prox} = 0.645 \pm 0.2$ km/s.

From the observed radial velocity values it would appear that Proxima is not gravitationally bound to α Cen AB. The problem, however, is that the entire issue of whether Proxima is gravita-

[56]This number (and the associated uncertainty) is taken directly from the research paper by the author, "Proxima Centauri: a transitional modified Newtonian dynamics controlled orbital candidate?" (*Monthly Notices of the Royal Astronomical Society*, **399**, L21, 2009).

tional bound to α Cen AB is (almost) entirely contained within the uncertainty of the radial velocity measurement deduced for Proxima. At the very best, at this stage, it can only be concluded that Proxima is just, or only marginally, bound to α Cen AB. It literally hovers on the divide between being α Cen C, the third star in the triple system with α Cen A and B, and Proxima, the star that just happens to be remarkably close to α Cen AB at the present time.

A study conducted by Jeremy Wertheimer and Gregory Laughlin (both at the University of California, Santa Cruz) put the question of Proxima's boundedness to the test by looking at the energy values associated with a series of cloned systems.[57] These systems were constructed by taking the observed values for system parameters and then randomly adding or subtracting terms within the range of the allowed observational uncertainty. A total of 10,000 clones were constructed, and it was found that about 44 % of such systems ended up having a negative total energy, indicating that Proxima was gravitationally bound to α Cen AB. The odds, at least from the presently available data, that Proxima can truly be designated α Cen C are currently no better than even.

Although this march-of-the-clones result is fair enough as it stands, alternative observational evidence suggests that Proxima really does form a trinity with α Cen A and B – that Proxima has the same estimated age and composition as α Cen A and B, and the sheer improbability that it would, when randomly observed, reside so close to α Cen A and B, all hint at a common origin. If one assumes that Proxima is indeed gravitationally bound to α Cen AB, then this sets very precise limits on the allowed radial velocity for Proxima, with $-22.3 < V(\text{km/s}) < -22.0$. The next step in answering the question "Is Proxima Centauri really equivalent to α Cen C?" will be entirely based upon obtaining a much more precise value for its radial velocity. Again, if one accepts that Proxima is gravitationally bound to α Cen AB, then what sort of orbit does it have? Wertheimer and Laughlin conclude that the orbit must be highly elliptical ($e \approx 0.9$) and have a major axis of order 2.6 pc (corresponding to $a = 272{,}212$ au). The orbital period would accordingly be of order 100 million years. Such an orbit could not possibly

[57] J. Wertheimer and G. Laughlin, "Are Proxima and Alpha Centauri Gravitationally Bound?" (*Astronomical Journal*, **132**, 1995, 2006).

be stable for more than a few cycles, however, and Proxima would soon be stripped from the gravitational grasp of α Cen AB. The size of the orbit can be much reduced (within the allowed uncertainties) if one adopts the argument that Proxima is most likely to be observed when it is close to apastron,[58] and in this case an orbit with a semi-major axis $a \sim 8{,}000$ au results, and the corresponding orbital period, comes down to about one million years.

The gravitationally bound status of Proxima is presently hidden within the uncertainty to which its radial velocity can be measured. Intriguingly, as well, a fundamental change in our understanding of the way in which gravity actually works might be hidden in the observational uncertainties. Early in the twentieth century, Einstein revolutionized the way in which we think about gravity – expressing it as an effect due to the curvature of spacetime – and he modified Isaac Newton's famous formula (as presented in the *Principia Mathematica,* first published in 1687) to account for the conditions of very high accelerations. For objects such as Proxima, which are moving in extremely low acceleration regimes, however, yet another change might come into play.

The idea of introducing a modified Newtonian dynamic (MOND) domain was first discussed within the context of galaxy rotation curves, and was presented as an alternative to postulating the existence (of the still mysterious) dark matter. Described in a series of fundamental research papers[59] published by Mordehai Milgrom (Weizmann Institute of Science, Israel) since the early 1980s, MOND relies on the postulate that in very low acceleration domains the way in which the gravitational force behaves changes. Indeed, Milgrom argues that once the acceleration acting on an object drops below a new fundamental (natural constant) value $a_0 \approx 1.2 \times 10^{-10}$ m/s^2, then its motion will become increasing different to that expected from the straightforward application of Newton's formula.

In effect, in the MOND domain, velocities should be higher than otherwise expected for the observed masses and separations.

---

[58] This expectation follows directly from Kepler's second law of planetary motion, which requires slower speeds and hence greater dwell times at apastron; the reverse situation applies at periastron.

[59] A good place to begin with respect to investigating the history and development of ideas pertaining to MOND is the website www.astro.umd.edu/~ssm/mond/.

For a standard two-body, Keplerian orbit with a small mass object orbiting around a larger central mass and interacting under a pure Newtonian gravitational interaction, the predicted orbital velocity V will decrease as the inverse square root of the orbital radius r: specifically $V^2 = GM_{AB}/r$. In this manner, the further distant the object is, the smaller will its orbital speed be.

For the same system in the domain where MOND applies, however, the orbital velocity will vary in an entirely different manner – namely as: $V^4 = a_0 GM_{AB}$. Indeed, in the MOND case the orbital velocity remains constant. (This, in fact, was the observed feature of galaxy rotation curves that resulted in MOND being developed.)

So, where does Proxima sit with respect to the pure Newtonian and MOND domains? Using the canonical values for M_{AB} and the observed separation distance of Proxima, the acceleration is $a_{prox} = 5.4 \times 10^{-11}$ m/s^2. Interestingly, therefore, it seems that $a_{prox} \approx \frac{1}{2} a_0$ and accordingly Proxima resides in the domain where MOND should be expected to apply. Indeed, the velocity of Proxima predicted by the MOND formula gives $V = 0.424 \pm 0.001$ km/s[60] – which is slightly larger than the standard Newtonian bound state limit ($V < 0.399 \pm 0.012$ km/s) but close to the lower limit allowed for the measured relative velocity of Proxima ($V = 0.645 \pm 0.2$ km/s). In the MOND situation, the orbit of Proxima around α Cen AB could be entirely circular, or, as found in a more detailed analysis,[61] it might have a slightly eccentric, $e = 0.2$, orbit with a semi-major axis of $a = 12{,}527$ au. A set of possible orbits for Proxima are shown in Fig. 2.28.

At this stage, nothing is for certain, and the entire question of Proxima's gravitationally bound status (that is, it being α Cen C) and the question concerning the existence of a MOND regime (and literally a new domain of gravitational physics) is hidden within the uncertainties that accompany the present radial

---

[60] This number (and the associated uncertainty) is taken directly from the research paper by the author, "Proxima Centauri: a transitional modified Newtonian dynamics controlled orbital candidate?" (*Monthly Notices of the Royal Astronomical Society*, **399**, L21, 2009).

[61] See the author's research paper, "The orbit of Proxima Centauri: a MOND versus standard Newtonian distinction" (*Astrophysics and Space Science*, **333**, 419, 2011).

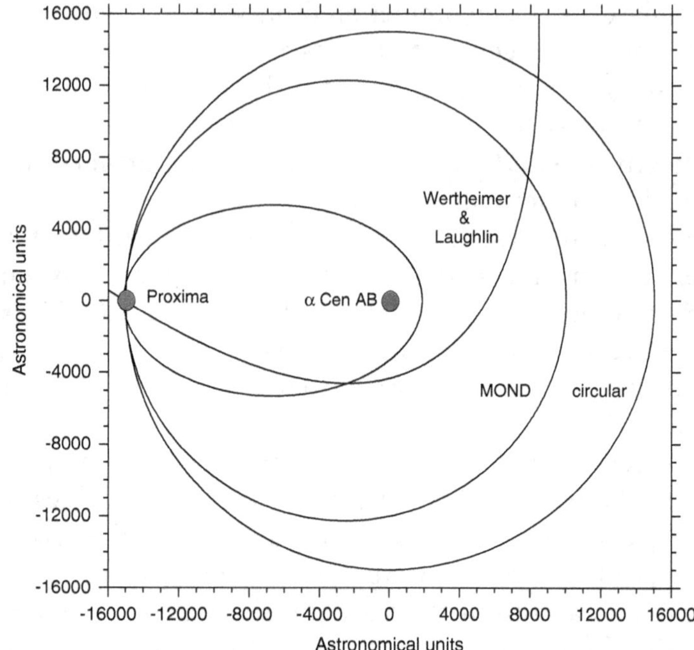

Fig. 2.28 Four possible orbits for Proxima around α Cen AB. The scale is given in astronomical units, and α Cen AB resides at the origin. The curve labeled *circular* is exactly that, and is the case where Proxima orbits α Cen AB at a fixed distance. The curve labeled *MOND* corresponds to the orbit having $a = 12{,}527$ au, $e = 0.2$. The curve labeled *Wertheimer and Laughlin* shows a fragment of the orbit deduced in Note 57. The smaller elliptical orbit is computed on the bases that Proxima is currently located at apastron, with respect to α Cen AB, and that it passes no closer to α Cen AB than the Hill sphere radius. (The Hill sphere radius, as introduced by American astronomer George William Hill in 1878, defines the limit interior to which the orbit of a smaller body will be significantly perturbed by the gravitational influence of the much heavier central body about which it orbits. In the case of Proxima the limitation is that its orbit is not significantly perturbed by passing too close to either α Cen A or α Cen B. The derivation for the Hill sphere radius used in Fig. 2.28 is given in Note 61.) The latter is taken as being the smallest possible elliptical orbit for Proxima, and it has an associated orbital period of 53,500 years

velocity measurements for Proxima. Reducing the uncertainty in the measurements is clearly a topic for future study and elucidation. For indeed, within its number resides the answer to two of the deeper and more carefully protected secrets of the α Centauri system.

3. What the Future Holds

Man over most of the world is in chains. Everywhere, powerful retrogressive forces are at work to keep him enslaved, or are fashioning new, more binding chains. All the powers of misused positivism are arrayed against him. But he will free himself if scientific knowledge can ever penetrate into his prison.

– A. E. van Vogt (Foreword to *Destination Universe*, 1952)

3.1 What Next?

The German physicist and pioneer of quantum mechanics Neils Bohr was once heard to remark that, "Prediction is difficult; especially if it is about the future." Indeed, the future is very difficult to pin down, and invariably the majority of futurists get it wrong. But, for all of its mercurial character, predicting the future is a little like looking out upon a fog-enshrouded landscape. Some features in the foreground and middle distance can be clearly seen. Even a far off mountain might just be visible, but much is obscured and no clear path to the emergent peaks can be discerned. All that is known for certain is that there is a road immediately before us, but how it branches and divides, and how it might weave and bend over in the long-run are completely unknown. We take a step, and then another and eventually arrive somewhere, but there is no guarantee that the arrival point is anywhere near where we were initially heading for. The path for each step ahead may have been fully illuminated and clear, but the end of the journey is not visible or even predictable from the starting position. We are in the realm of the ultimate butterfly effect, where even the smallest,

M. Beech, *Alpha Centauri: Unveiling the Secrets of Our Nearest Stellar Neighbor*, Astronomers' Universe, DOI 10.1007/978-3-319-09372-7_3,

most minute of changes in cadence and/or direction will have dramatic and initially unknowable consequences.

For all of our uncorrectable myopia concerning the deep future, however, we may nonetheless see and even successfully predict some of what might come to pass in times yet to come.

Writing during the ascendancy of the Cold War years, A. E. van Vogt, as indicated in the somewhat dire opening quotation to this section, felt that the future of humanity would be purely determined through scientific innovation. This, of course, is only partially true. Freedom is not so easily achieved or corralled. Certainly, technological advancement will facilitate the means of future development and exploration, but it will be human grit, imagination and social evolution that will drive the pace.

In this final section it is my intention to cast a broad net, a net encompassing many diverse oceans of thought, certainties and speculations. The near-term future, as far as astronomical goals are concerned, is clearly discernible, and our net will have a small weave. Moving deeper into the future, however, a century or so from the present, the current desires to develop interstellar travel are clear enough, but the means of facilitating its realization are highly uncertain. Our net, accordingly, must have a large weave and may even sport gaping holes, and much of what might seem certain now may never come to pass.

Moving into the ultra-deep future, the ultimate activities and achievements of humanity fade into obscurity and may literally go anywhere. Oddly, however, as our view of what our distant descendants might achieve dissolves into impenetrable noise, our understanding of what will happen to the Sun and the stars in the solar neighborhood move into sharp focus. Indeed, we can say now, with a considerable degree of certainty, what the Sun, α Centauri, Barnard's star and even Teegarden's star will be like a billion, two billion, even ten billion years hence. The future beckons, and it promises great adventure.

3.2 More Planets?

There is little doubt that during the next several decades the primary reasons that astronomers will continue to study α Centauri are (i) to confirm, or indeed, refute, the existence of α Cen Bb, and (ii) to identify the total number of planets within the system. As already seen, there is every reason to suspect that there are more planets, given the confirmation of α Cen Bb, in orbit around α Cen B, as well as potentially multiple numbers of planets in orbit around α Cen A as well as Proxima.

As the future continues to tick forward astronomers, in addition to looking for Centaurian planets, will also continue to deep-probe the solar neighborhood for new diminutive dwarf stars and brown dwarfs – objects hovering on the dynamic edge of present generation technological and threshold detection. On a smaller physical scale, astronomers will additionally maintain the search for rogue planets and interstellar comets[1] – bodies gravitationally cast adrift from their natal stars. Indeed, while the August Douglas Adams, author of *The Hitchhikers Guide to the Galaxy* series of books, reminded us that space is big, it is not devoid of substance, being, as it is, infused with a dynamic litany of diverse astronomical objects.

The distance between the Solar System and the nearest free-floating rogue Jupiter is not known, but data provided by the 2010–2011 Wide-field Infrared Survey Explorer (WISE) spacecraft has been deep searched to reveal that no brown dwarfs are likely to reside within or even close to the outer domain of the Solar System. Not only this, the WISE spacecraft data also rule out the possibility of any additional Saturn-mass or larger planets existing within 10,000 au of the Sun. There are no objects with a mass comparable to Jupiter within 26,000 au.[2]

---

[1] Numerous rogue, or free-floating, planets have now been detected, but it is still far from clear whether they must all have been formed and then ejected from the disks of newly developing planetary systems, or whether some might have formed by direct gravitational collapse. No interstellar comet has ever been unambiguously recorded. The best indicator of such an origin would be the discovery of a comet on a highly hyperbolic orbit (that is, a comet having an eccentricity well in excess of 1). This being said, the number of interstellar cometary nuclei is likely to be extremely large.

[2] These results effectively rule out the once popular idea of a large planet/brown dwarf object, variously named Planet-X, Nemesis and Tyche, on a highly eccentric orbit, periodically passing through the Oort Cloud region and thereby triggering cometary showers and planetary-cratering epochs within the inner Solar System. Although an astronomical trigger seems less likely in light of the WISE survey, the problem of explaining the apparent 25- to 30-million-year periodicity in terrestrial impact crater ages remains unsolved. The WISE survey data does not rule out the possibility of a terrestrial mass planet existing within the Kuiper Belt region. Indeed, one of the most compelling theories for the origin of the Kuiper cliff – where the number of observed classical Kuiper Belt objects drops precipitously, at a distance of about 50 au – is that of invoking gravitational clearing by a Mars to Earth-sized planet.

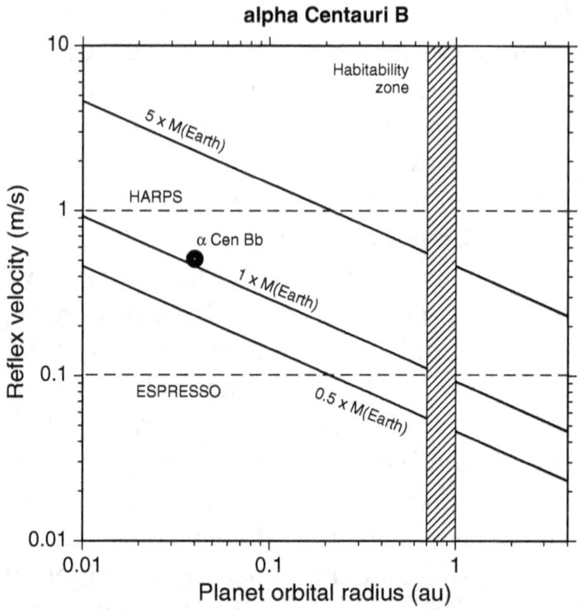

FIG. 3.1 The reflex velocity of α Cen B for planets within the stability zone having masses 0.5, 1 and 5 times that of the Sun. Here we assume that the planetary orbits are seen under the edge-on, best viewing geometry (recall Fig. 2.21). For the same orbital range and planetary masses, the reflex velocity for α Cen A will be nearly identical to those shown here. The habitability zone, however, is located further outwards between 1.17 and 1.73 au. The horizontal lines labeled HARPS and ESPRESSO indicate the present detection limit for the High-Accuracy Radial velocity Planetary Searcher and the design detection limit for the Echelle Spectrograph for Rocky Exoplanets and Stable Spectroscopic Observations instrument

In terms of detecting planets around nearby stars, and especially α Cen A and B, the future looks particularly promising. In terms of radial velocity detection surveys the requirements for next generation designs are clear enough, and illustrated for α Cen B in Fig. 3.1. With the present-day detection limit of 1 m/s, a 5-Earth mass planet would have to be located closer than 0.2 au to α Cen B in order to be clearly detected. Conversely, to detect an Earth mass planet in the habitability zone situated around α Cen B will require an ability to measure a reflex velocity to of order 10 cm/s (or better).

As revealed earlier, the current state of the art, the HARPS instrument's (recall Fig. 2.15), limit for reflex velocity measurements is of order 1 m/s. The ESPRESSO instrument, currently under development by engineers at the European Southern Observatory and scheduled for first light in 2016, will push this limit an order of magnitude smaller, to some 10 cm/s. With appropriate high cadence time data collection and statistical binning, this limit may well be pushed further downwards by a factor of two or three. Change is upon us, and the rate of technological advancement is impressive. Thirty-five years ago the precision of radial velocity measurements was of order tens of meters per second; in the very near future it will be at the level of several centimeters per second.

Planet transit searching will also continue to find new systems as we move through the next several decades. Following in the pioneering path established by the French Space Agency's CoRoT and NASA's *Kepler* missions, the European Space Agency (ESA) has recently announced plans to develop the CHaracterising ExOPlanets Satellite (CHEOPS) mission. Set for launch in 2017, this mission will monitor solar neighborhood stars for brightness variations indicating the presence of transiting exoplanets. The mission design will enable the detection of planets in the super-Earth to Neptune-size range, and the spacecraft will also carry instruments to enable the study of planetary atmospheres. In the near term, the recently launched Gaia[3] satellite from ESA will begin to return results on new planetary systems following the completion of its 5-year survey mission in 2018. This specific spacecraft will use astrometric methods to find exoplanets, and, indeed, the mission objectives state that "Every Jupiter-sized planet with an orbital period between 1.5 and 9 years" will be discovered out to a limiting distance of 150 light years (50 pc).

Scheduled for a launch date in 2017 is the NASA-driven Transiting Exoplanet Survey Satellite (TESS) mission. This project is essentially a next generation Kepler satellite survey mission to detect transiting planets around nearby stars. Rather than having a single staring mode, however, TESS will steadily monitor the entire sky during its planned 2-year mission lifetime. Its primary goal is

[3] In the modern fashion of ignoring conventional language usage, the GAIA mission originally took its name from the acronym for Global Astrometric Interferometer for Astrophysics. Later mission design changes, however, resulted in a revised observational technique being employed, and while the acronym is no longer accurate, the mission name has remained the same.

to identify Earth and super-Earth-sized planets in orbit around some 500,000 stars, in the F to M spectral-type range, located within 50 pc of the Sun.

With respect to the discussion that will follow below, it is worth noting here that the TESS mission grew out of a projected developed at MIT in 2006 with seed funding from the Internet company Google. The initial plan was to design and operate a privately funded space mission. The fact that the original plan did not succeed speaks volumes, as we further discuss below, about our current collective inability to organize and successfully complete industry-funded, not-for-profit, research.

Working to a slightly longer timescale than both CHEOPS and TESS is the ESA-funded Planetary Transits and Oscillations of Stars (PLATO) mission. Like the CoRoT, Kepler CHEOPS and TESS missions, the PLATO spacecraft will search for signs of exoplanet transits around nearby stars. Scheduled for launch between 2022 and 2024, the provisional PLATO spacecraft design calls for the integration of 34 telescopes and cameras that will continuously monitor some one million stars spread across the entire sky during a mission timescale of at least 6 years. The key goal of the PLATO mission is to detect Earth-sized exoplanets located within the habitability zones of their parent stars.

In addition to dedicated transit detection spacecraft, the launch of the James Webb Space Telescope in late 2018, as well as the possible funding of new spacecraft missions utilizing advanced imaging techniques,[4] will greatly enhance the ability of astronomers to not only find new exoplanets but also subject their atmospheres to highly detailed analysis. Indeed, next to finding more planets in the future, one of the primary goals of astrobiology will be to find planets located within the parent star's habitability zone. For such planets, high resolution spectroscopy techniques should be able to determine the basic atmospheric composition as well as deep-search them for possible biomarkers. Indeed, future space and ground-based telescope research will seek to measure the composition, temperature and reflectance (albedo) properties

[4] Such missions might employ, for example, interferometer techniques and multiple satellite systems – such as that proposed in NASA's (now canceled) Terrestrial Planet Finder (TPF) mission.

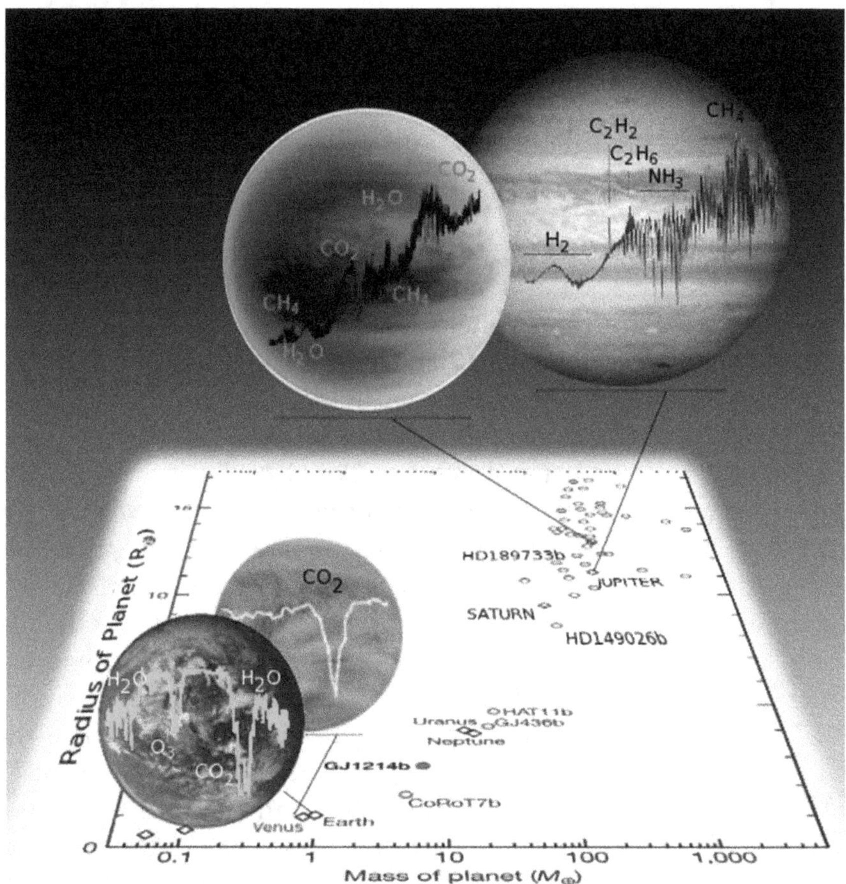

FIG. 3.2 Planets of the same physical size can have very different atmospheric properties, according to where they are located within their parent system. In our Solar System, for example, Earth and Venus, although almost identical in size and mass, have very different atmospheric structures (the former showing biomarkers such as O_2, O_3 and H_2O, the latter showing mostly CO_2). Likewise, Jupiter in our Solar System and the hot Jupiter HD 189733b have similar sizes, but the latter, being much closer to its parent star, shows distinctly different atmospheric properties (Image courtesy of ESA and the EChO consortium)

of exoplanet atmospheres. Such studies will seek to identify spectral characteristics (Fig. 3.2) relating to molecules such as CO, CO_2, CH_4, NH_3 as well as key biosignature molecules such as H_2O and O_3.

Early results in this direction, obtained with the Hubble Space Telescope along with the Spitzer infrared telescope, have already allowed the clear identification of such molecules as CO and H_2O in the atmospheres of several hot Jupiters. Indeed, Hubble and Spitzer observations of HD 189733b, an exoplanet discovered in 2005, reveal the presence of H_2O, CH_4 and CO within its atmosphere. Additionally, the infrared Spitzer spacecraft data has enabled the construction of a thermal map showing the temperature variation across the planet's atmosphere – the day/night subpoint temperatures being 1,212 and 973 K respectively.

The CO detection within the atmosphere of HD 189733b was actually made from the ground with the European Southern Observatory's 8.2-m Very Large Telescope (VLT) in Chile, and indeed, such studies are likely to become more commonplace over the next decade. The key instrument used in the VLT study was the Cryogenic InfraRed Echelle Spectrograph (CRIRES) – the first prototype assembly seeing operation in 2006 and being specifically designed to provide extremely high resolution spectra. The next generation instrument CRIRES + is scheduled to come online in 2017, and combining similar such instrumentation with the planned 39.3-m European Extremely Large Telescope (E-ELT), scheduled for first light in the mid-2020s, will result in a dramatic increase in diagnostic ability. Indeed, it has been determined that the E-ELT, with a state-of-the-art high resolution spectrograph, should be able to detect the infrared O_2 biomarker,[5] at a wavelength of 0.76 μm, in an Earth-like planet located within the habitability zone of an M-dwarf star – indeed, a dwarf star just like Proxima Centauri.

Although the presently funded projects for exoplanet detection and study, both ground- and space-based, are predicated on the development or extension of well-known engineering principles, it is highly likely, if the history of science is to be any guide, that the next really momentous step in furthering our collective astronomical reach will come from technologies hardly realized in the present epoch. One such innovation might be the development of the quantum telescope as recently described by Aglaé Kellerer

[5] This, recall from Sect. 2.16, is one of Carl Sagan's "criteria for life" indicators – the existence of gaseous O_2 being taken as an indicator of biotic activity.

(Durham University, England). In this case the properties of entangled photons are used to push the resolution limit of a telescope well below its classical value. The ability of a telescope to reveal fine-scale detail is related to its diffraction or resolution limit *R*. This limit is set according to the wavelength λ of light being studied and the diameter *D* of the telescope objective – specifically $R \sim \lambda/D$. At a fixed wavelength the only way to improve image resolution, that is, make *R* smaller, is to built a bigger diameter telescope, and this, of course, has a cost overhead.

It is generally assumed that the cost of building a new telescope varies as the diameter *D* of the telescope raised to the power 2.7. That is, $cost \approx K\ D^{2.7}$, where the constant $K \approx 3 \times 10^5$. Accordingly, a 10-m class telescope, such as the Keck on Mauna Kea in Hawaii, or the Gran Telescopio Canarias on La Palma, costs about $150 million to build and house. The massive, next-generation, 40-m E-ELT is scheduled to cost about $1.5 billion (the standard formula suggests it might cost more like $6 billion).

The E-ELT will have a light gathering power 16 times greater than that of, say, the Gran Telescopio Canarias, but its resolution will only be 4 times better. To improve upon the resolution of a 10-m sized telescope by a factor of 10, a telescope with a 100-m diameter mirror would need to be constructed, and this not only pushes the extreme limits of engineering possibilities, it would also be prohibitively expensive – our simple formula suggesting a cost of something like $75 billion (interestingly, this amount of money is about the same as the total annual budget for all of the various U. S. intelligence-gathering agencies). It would seem likely therefore that the limits of new telescope funding will be reached once *D* achieves the 40- to 50-m mark.

This budgeting limitation on construction brings us back to the recent work of Aglaé Kellerer, who argues that by being clever with the light received at the telescope, the resolution limit *R* could in principle be pushed well below its classical value. The trick, Kellerer argues, is to clone the photons from the incoming astrophysical light source.

Moving beyond the simple act of amplification, however, the technological challenge will be to clone and quantum entangle the incoming starlight. Although the idea of quantum entanglement caused Einstein to interject his "spooky action at a distance" outburst, modern research shows that if N photons are entangled then the diffraction limit of the ensemble is such that $R \sim \lambda/ND$. There are additional complications that we do not follow here, but Kellerer shows that if the light from an astronomical source entering a telescope tube is passed through an appropriate amplifying medium, that generates N entangled photons for each incoming photon, and if one only records the detail in the cloned and entangled photons, then the resolution limit is reduced by a factor relating to the square root of N. The important point to remember at this stage is that the improvement in resolution has not required any change in the diameter of the recoding telescope. Indeed, if N can be made as high as 100, then a ten times increase in resolution can be achieved. The point now is that one does not need to build a bigger telescope to record more fine-scale detail. Although the era of the practical quantum telescope is still likely 25–50 years away, the very possibility of their construction highlights the point that bigger is not always better and that what really counts is being smarter.

The near-term future of exoplanetary science will, with little doubt, provide us with new and exciting discoveries. The conditions under which we should move forward are reasonably well understood, and the challenge to realize the development and eventual deployment of new instruments and new technologies has already been taken up. Time, engineering skill, persistence, imagination and data gathering will, for the next few decades, dictate the pace of discovery.

3.3 A Stopped Clock

Perhaps one of the most wonderful examples of a convergence between purely speculative (and indeed, satirical) literature and reality is that penned by Jonathan Swift in his *Travels into Several Remote Nations of the World* – better known as *Gulliver's Travels* – first published in 1726. In describing the achievements of the astronomers located aboard the flying island of Laputa, the narrator, Lemuel Gulliver, noted that two moons had been discovered to orbit the planet Mars. Mars, as we now know, does indeed have two diminutive moons, but they were not discovered until

American astronomer Asaph Hall swept them up in 1877, some 141 years after Swift's text first saw print.[6]

There is no real logic behind Swift's choice of two moons for Mars – although it has been suggested that since Venus has no moons, Earth has one, then logically Mars should have two. But by this same logic we are presumably led to the conclusion that Mercury should have minus-one moons! Clearly, the number chosen by Swift was immaterial to his narrative, and it just happened to be the right number with respect to reality.

Just for fun, however, let us continue the theme set in motion by Swift, and upon the basis that even a stopped clock (a mechanical analog one, that is) tells the correct time twice a day, let us briefly review, for future posterity, how many planets and moons have been invoked within a (small) sample of science fiction works for the stars of α Centauri.

The results of our study are shown in Table 3.1, and they are not, as might be imagined, particularly surprising or enlightening. For α Cen A and B the number of planets has varied between zero and four. An interesting mixture of Jovian and terrestrial-sized planets has been invoked by our sample of authors. Invariably, or nearly so, at least one habitable terrestrial planet or moon is additionally detailed, and this, of course, simply reflects the fact that the various narratives must have somewhere for the human adventurers to visit and/or live upon.

The first author to suggest that α Cen A might have planets appears to be Friedrich Mader in his 1932 adventure novel *Distant Worlds*. Although Mader has been described as "the German Jules Verne," his ripping-yarn space-travel novel, while absolutely charming, stretches, at times, one's incredulity to near breaking-

---

[6] Swift writes, in fact, that the moons have orbital radii of 3 and 5 Martian diameters, with periods of 10 and 21.5 h. The actual orbital radii for Phobos and Deimos are 1.38 and 3.45 Martian diameters, with their periods being 7.656 and 30.288 h, respectively. The two moons are in fact diminutive in size, with Phobos having a mean diameter of just 22 km, while that for Deimos is 12 km. It is not presently clear if the moons were captured after Mars formed, or if they are original, planetesimal objects that have always orbited the planet. Though the orbital values deduced and attributed to the Laputan astronomers are remarkable for their closeness to reality, it would have been helpful if they had also recorded that the orbits are in a state of slow change. Indeed, Phobos is gradually getting closer to Mars, and it will break apart due to gravitational tidal forces and/or crash onto the planet's surface some several tens of millions of years hence. Deimos, in contrast, is gradually moving further away from Mars.

TABLE 3.1 The number of Centaurian planets and associated moons as described in a small sample of science-fiction works, movies and video-games published between 1932 and 2009. HZ indicates that the planet was deemed to be in a habitable region of the star system. (i) Number not specified. (ii) Brackett appears to be unaware that α Centauri is not a single star. (iii) This is actually a serious and detailed science article, rather than a science-fiction work, published in the *Journal of the British Interplanetary Society*, **29**, 611–632 (1976). (iv) One of the planets is an Earth-sized waterworld developed and maintained through terraforming. (v) Several of the moons have atmospheres and liquid water upon their surfaces. The two outermost moons have retrograde orbits suggestive of being capture asteroids/Kuiper Belt like objects. The second Jovian planet has 2 planetoid companions at the L4 and L5 Lagrange points

α Cen A	α Cen B	Proxima	Work, author and year published
1 terrestrial (in HZ) + 3 moons	(i)	(i)	*Distant Worlds*, F. Mader, 1932
(i)	(i)	6 planets (1 in HZ) + ring system	*Proxima Centauri*, M. Leinster, 1935
4 (all in HZ)	4 (all in HZ)	0	*Far Centaurus*, A. van Vogt, 1944
0	0	0	*Seed of Light*, E. Cooper, 1959
3 planets (1 in HZ) (ii)	(i)	(i)	*Alpha Centauri or Die!*, L. Brackett, 1963
Yes (i)	Yes (i)	4	*A Program for Interstellar exploration*, R. L. Forward, 1976 (iii)
4 terrestrial (iv) + Asteroid Belt	5 terrestrial	(i)	*Foundation and Earth*, I. Asimov, 1986
1	(i)	(i)	*The Songs of Distant Earth*, A. C. Clarke, 1986
5	3	3 planets, one with a moon	*Flying to Valhalla*, C. Pellegrino, 1993
4	4	"handful of Ceres-like objects"	*Alpha Centauri*, W. Barton & M. Capobianco, 1997
1 terrestrial (in HZ) + 2 moons	(i)	(i)	*Sid Meier's Alpha Centauri*, M. Ely, 2000

(continued)

(continued)

α Cen A	α Cen B	Proxima	Work, author and year published
1 Jovian + moon 1 terrestrial	1 (in HZ)	0	*First Ark to Alpha Centauri*, A. Ahad, 2005
1 Jovian + 6 moons 1 terrestrial	1 (in HZ)	8	*Dangerous Voyage to Alpha Centauri*, F. Reichert, 2007
3 Jovian, one with 14 moons (v)	(i)	(i)	*Avatar*, James Cameron, 2009
1 terrestrial (in HZ)	(i)	1 (in HZ)	*Rigel Kentaurus*, Rick Novy, 2012
yes (i)	yes (i)	6 terrestrial (1 in HZ) + Kuiper Belt	*Proxima*, Stephen Baxter, 2013

point (and this was probably so even for 1930s audiences). Nonetheless, after a whistle-stop tour of all the Solar System's planets (Pluto surprisingly excluded), the Asteroid Belt and several comets, the happy crew arrive (by way of an interstellar comet traveling at 50 times the speed of light!) arrive at the planet Eden. The planet is described as having a circumference twice that of Earth[7] and a rotation period of 50 h. The planet supports a breathable atmosphere, an ocean system, a diverse biosphere and is inhabited by an advanced, utopian civilization. It also has three (different-colored) moons, one of which has a retrograde orbit. It is of interest to note that on their return journey to Earth Mader's adventurers encounter a 'dark star' that, from its description, is really a lone planet in interstellar space. In this manner Mader effectively anticipates the equivalents of exoplanet super-Earths and free-floating, rogue Jupiter planets.

Somewhat surprisingly the number and types of planets suggested as possible companions to Proxima Centauri has varied more than those invoked for α Cen A or B – indeed, for Proxima

[7] This additionally makes Eden two times larger than Earth, and given a similar bulk density to that of Earth it would make it eight times more massive – a super-Earth by modern standards.

the number of planets has varied between zero and eight. Again, surprisingly, for all of the contemporary astronomical facts and snippets that Mader drops, Jules Verne-like, into his *Distant Worlds* narrative, no mention of Proxima is even made.

The first author to invoke planets around Proxima seems to have been Murray Leinster, in his 1935 short story simply called "Proxima Centauri." Writing 20 years after Proxima's discovery, Leinster invokes within his story the presence of six Earth-like planets as well as a distinctive dust ring – "Quant, that ring. It is double, like Saturn's," comments one of the novel's characters. One of Leinster's hypothetical planets is capable of supporting human life, while another is home to an ambulatory, carnivorous plant-like creature. The humans completely destroy the latter planet after a series of fatal encounters between the humans and the Venus flytrap-like Centaurans.

Other imagined planets set in motion around Proxima are described as being dwarf planets or asteroids, like Ceres in our own planetary system; some of the planets are deemed to be habitable, and some have at least one moon.

In addition to knowing that no large, close-in Jovian planets exist within the α Centauri system, it is additionally known that α Cen B has at least one Earth-sized planet outside of the habitability zone, and these observations begin to significantly narrow down the posthumous prize for serendipitously predicting the number of planets within the α Centauri system. Such is the hard life of the would-be futurist, and only time and many extensive observational campaigns will tell how reality measures up to human guesswork.

3.4 Planets Aside: Comets and Asteroids

"In bright ellipses their reluctant course; orbs wheel in orbs, round centres centres roll, and form, self-balanced, one revolving whole." So writes Erasmus Darwin in his 1791 pastoral "Botanic Garden." These lines remind us that there is more, indeed, very much more to a planetary system than its planets. Earth, in its yearly rounding of the Sun, testifies to this very notion as it ploughs through a miasma of meteoritic dust particles and bolder-sized rocks, which upon contact, for a breathless moment, brighten the sky as meteors and fireballs, and occasionally as meteorite-dropping bolides (the literal *Draco Volans* of the night sky).

This particulate sea through which Earth annually pushes is the tip of the proverbial iceberg, and the presence of the dust and boulders betrays the existence of extensive reservoirs of asteroids and cometary nuclei. Most of the asteroids orbit the Sun between Mars and Jupiter, although deviant members can escape the main-belt precinct to prowl the inner Solar System – lurking as potentially Earth-impacting objects. The asteroids themselves are ancient; indeed, they are the remnant originals, solid bodies of rock and iron that avoided the press of assimilation into planetary interiors. Ancient first-born, the asteroids have slowly been grinding themselves, through mutual collisions, into smaller and smaller structures, and it is the smallest shards ejected in these collisions that occasionally fall to the ground to be collected by startled humans as meteorites.

Does α Centauri have an Asteroid Belt, and might meteorites be collectible on Centaurian planets? The answer to these two questions is yes and possibly. Certainly there is every reason to suspect that asteroid-like objects exist in orbit around both α Cen A and B irrespective of the formation of any larger planetary structures. And, assuredly, if there are terrestrial planets supporting atmospheres in orbit around either or both α Cen A and B, then Centaurian meteorites will also occur.

In contrast to the situation for Earth in our Solar System, however, it is highly unlikely that a Centaurian planet will experience periodic meteor showers. Earth experiences such displays when it passes through the debris trail left in the wake of a comet. The dust grains, originally embedded within the cometary ice, spread around the comet's orbit, and if the geometry is right then once, and sometimes twice, per year Earth can sweep up the grains (which are destroyed as meteors in the upper atmosphere) in its passage through the meteoroid stream. Meteor showers do not last forever, however, and through gravitational scattering, solar radiation pressure and collisional break up a stream is gradually dispersed.

Although each time the parent comet rounds perihelion it feeds more material into its associated meteoroid stream, comets themselves have a finite activity lifetime. In perhaps a few millions of years the entire show is over with, the parent comet will have become inactive and devolatized, possibly even vaporized in the Sun, and its meteoroids will have been destroyed in a brief blaze of atmospheric glory, or scattered deep into the Solar System, to form part of the sporadic background.

Long by human standards, the few-million-year lifetime of a meteoroid stream is short compared to the age of the Solar System (4.56 billion years), and the very fact that we see meteor showers to this very day tells us that there must be a long-lived reservoir of cometary nuclei continuously feeding into the family of comets capable of producing meteoroid streams. In our Solar System there are in fact two reservoirs of cometary nuclei[8],[9]: the Kuiper Belt and the Oort Cloud.

Figure 3.3 provides a schematic diagram of the extreme orbital points of the various cometary families and cometary reservoirs within the Solar System. The diagram plots greatest (aphelion) distance from the Sun against the closest approach (perihelion) distance. In the upper left-hand corner we find the first cometary reservoir: the Oort Cloud, which we discuss more fully below. The second reservoir resides in the Kuiper Belt region beyond the orbit of Neptune. It is the gradual leaking out and migration of cometary nuclei from these two reservoirs that feeds into the potential meteor shower-producing comet families. Collisions between Kuiper Belt objects feed fragments into the inner Solar System along the ecliptic plane,[10] ultimately to be captured (or destroyed, as in the case of Comet Shoemaker-Levy 9 in 1994) when they encounter Jupiter. It is the Jupiter family of comets that mostly provide Earth with its annual meteor showers.

Cometary nuclei are also fed into the inner Solar System from the Oort Cloud, and these nuclei populate the long period, seen only once or very, very rarely in the inner Solar System, family of comets (one recent such example being comet C/21012S1 – ISON). These comets have large aphelion distances but potentially very

[8] There is good recent evidence to indicate the existence of a third cometary reservoir concomitant with the main-belt asteroid region.

[9] See, for example, the excellent website hosted by David Jewitt (UCLA): http://ww2.ess.ucla.edu/~jewitt/David_jewitt.html.

[10] This is the plane of Earth's orbit around the Sun.

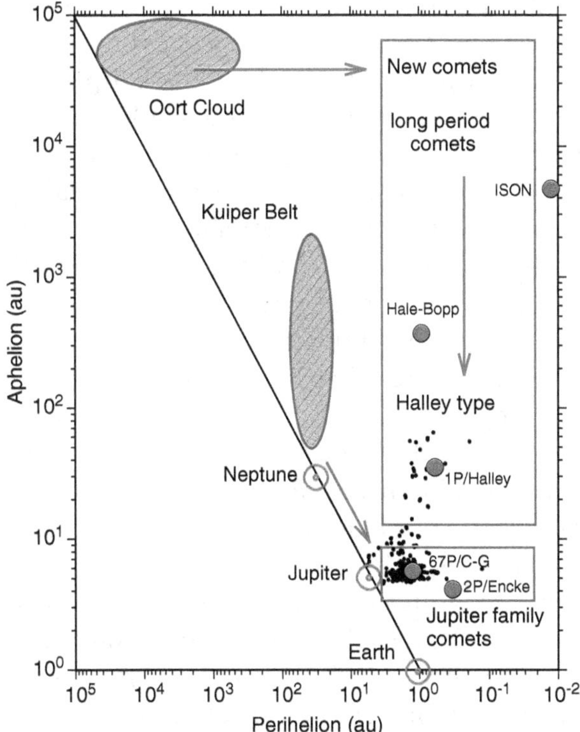

FIG. 3.3 Range of aphelion versus perihelion distances for the cometary families and reservoirs (cross-hatched) in our Solar System. The small dots correspond to a selection of known Jupiter family and Halley-group comets. Also indicated are two well known long-period comets: Comet ISON (which disintegrated during its 2013 perihelion passage) and Ccomet Hale-Bopp (discovered in 1995). The location of the first known periodic comet, 1P/Halley, is indicated, and this is the archetype of the Halley group of comets. The location of the second known periodic comet, 2P/Encke, along with the 2014 target nucleus, 67P/Churyumov-Gerasimenko (67P/C-G), of ESA's Rosetta spacecraft are indicated, and these objects fall within the region of the Jupiter family of comets. The scale is in astronomical units and the diagonal line corresponds to circular orbits for which the perihelion and aphelion distances are equal

small perihelion distances (0.012 au in the case of Comet ISON). If a long-period comet chances to pass close to one of the outer planets, and especially massive Jupiter, then it might be gravitationally captured into a shorter period orbit, and this process populates the so-called Halley group of comets.

Due to the nature of the Oort Cloud, the long period and Halley family of comets can approach the inner Solar System at any angle to the ecliptic (not just along the ecliptic, as in the case of Kuiper Belt-derived cometary nuclei). Halley's Comet exemplifies this situation in that its orbit is retrograde and inclined by 162.3° to the ecliptic. The orbital orientation of Halley's Comet is such that it can produce two meteor showers on Earth each year: one in May – the Eta Aquarids – and one in October – the Orionids.

In our Solar System, meteor showers are essentially possible because of the existence of Jupiter. It is the great mass and gravitational influence of Jupiter that both deflects and traps (albeit for just a short while) cometary nuclei into families that can then go on to produce meteor showers on the inner planets (indeed, along with Earth, Mars and Venus, have their own set of meteor showers). Such conditions will not exist for any terrestrial planets in the α Centauri system. First, there are no Jovian-mass planets to trap and control the dynamics of cometary families, and second there is no analog of the Kuiper Belt to feed cometary nuclei into the inner Centaurian region. All this being said, meteor storms, rather than periodic meteor showers, could yet be an observable phenomenon in the skies above any planets (with an atmosphere) in the α Centauri system.

However, as we shall see below, there is every reason to believe that a cometary reservoir similar to our Oort Cloud should exist around α Cen AB, and this will occasionally feed cometary nuclei into the regions close to either one of stars. It will be during such sundives that our putative planet might encounter a cometary dust stream and thereby experience a one-off meteor storm – the sky literally lighting up with thousands of shooting stars.

Meteor storms also occur in our Solar System. The November Leonids, during the late 1990s, for example, provided several meteor storms in our Earthly sky. More recently, in October of 2014, the long-period comet C/2013 A1 Siding Spring produced a meteor storm in the atmosphere of Mars.

The Oort Cloud delineates the boundary of our Solar System. Its outer edge indicates where the Sun's gravitational influence ends and where interstellar space finally begins. Although not perfectly spherical in shape, the Oort Cloud has a radius of about 100,000 au and literally stretches halfway to α Centauri.

Although there are still a few dissenting voices, the general consensus is that Oort Cloud cometary nuclei did not form in the region where they are currently found. Rather, they formed much closer in towards the Sun, in the same region, in fact, of the solar nebula where the Jovian planets themselves formed (at and beyond the ice line). Gravitationally stirred, shifted and ejected from their original orbits many of the original cometary nuclei were flung into interstellar space. Others, however, where the gravitational slingshot effect was less dramatic, were able to cling to the Solar System, entering into orbits that ultimately populated the Oort Cloud region. Although the initial ejection of the cometary nuclei would have been in the ecliptic plane, the random passage of nearby stars, and the gravitational tides of the Milky Way Galaxy at large, have stirred up the Oort Cloud cometary orbit inclinations so that they now exhibit an isotropic distribution.

There is no specific reason to suppose that the formation of a large, boundary-defining cloud of cometary nuclei is unique to the Solar System, and it is to be expected that all Sun-like stars will have, at some level, similar structures to our Oort Cloud. Certainly, the cometary nuclei (as well as rock/iron asteroid bodies) are expected to accompany the formation of all low mass stars irrespective of formation of any actual planets. Indeed, recent surveys at both optical and infrared wavelengths have found numerous stars supporting extensive disks of dust derived from cometary and asteroid collisions (Fig. 3.4).

No distinctive dust cloud has, to date, been detected around α Cen AB. This does not mean, of course, that there is no cometary cloud. Although the cometary cloud around α Cen AB might not be dense enough to generate a distinctive dust disk,[11] a swarm of cometary nuclei may yet be detectable through the remote monitoring of Proxima. Here the idea is to look for flares induced by cometary impacts.[12] The problem, however, is that, as we have already seen, Proxima is a known flare star, and not all of the flares

---

[11] The most comprehensive study to date is that by J. Wiegert et al., "How dusty is α Centauri? Excess or non-excess over the infrared photospheres of main-sequence stars" (*Astronomy and Astrophysics*, **563**, id. A102, 2014).

[12] The author first discussed this idea in the article, "Exploring α Centauri: from planets, to a cometary cloud, and impact flares on Proxima" (*The Observatory*, **131**, 212, 2012).

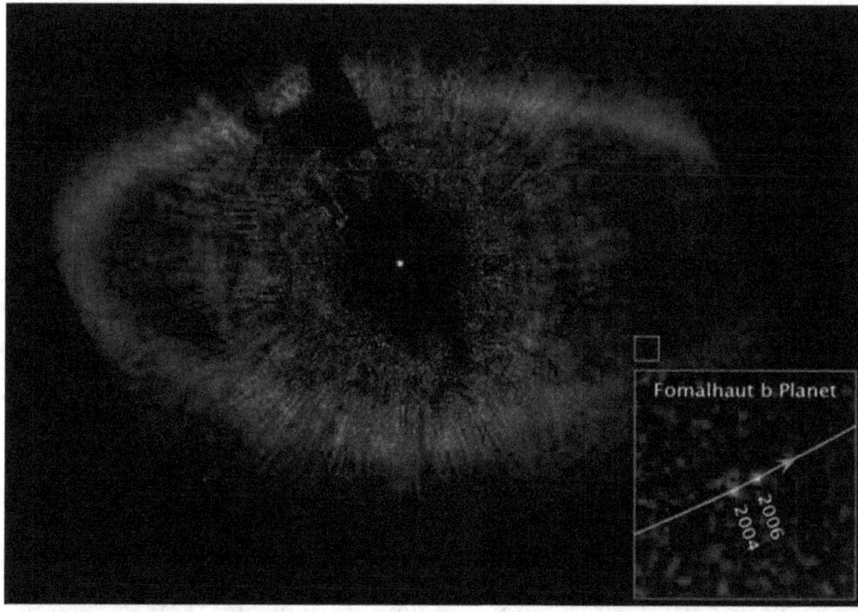

FIG. 3.4 The extensive dust disk around the star Formalhaut (α Piscis Austrini). Formahaut is at the image center, but is covered by an artificial occultation plate. The dust found within this system is derived through the collision of numerous cometary nuclei. Indeed, to maintain the disk against radiation driven mass-loss the amount of dust that must be produced per day is equivalent to the destruction of some 2,000 1-km sized cometary nuclei. The disk is some 200 au across. The inset shows the motion of Formalhaut b, a cold Jupiter planet that has been perturbed from its initial formation location into a highly elliptical orbit that carries it well away from Formalhaut (Image courtesy of ESA and the Herschel Space Observatory)

will be due to impacts. This being said, an estimate for the number of cometary impacts that might occur on Proxima can be made by taking our Oort Cloud as being typical. Indeed, following a detailed study[13] of formation dynamics J. G. Hills (Jet Propulsion Laboratory) has argued that cometary clouds should be present in all planetary systems that support at least one massive Jovian planet, and/or around binary stars. It is the latter condition that applies to the α Centauri system, and it is the binary parent condition that is

[13] J. G. Hills, "Comet Showers and the steady-state infall of comets from the Oort cloud" (*Astronomical Journal*, 86, 1730, 1981).

needed to boost the energy and change the orbits of cometary nuclei, so that they can move from their formation locations (near the ice line) into the remote halo.

Hills also argued that the sizes of cometary clouds must be about the same, independent of the host star system, since it is predominantly the gravitational influence of the galactic gravitational field that sets the outer boundary location at about 100,000 au.[14] Additionally, the inner boundary of the cometary cloud should also fall, independent of the parent system, at about 20,000 au. The inner boundary of a cometary cloud is not a physical boundary, as such, but rather it defines the region of relative orbital stability against the gravitational perturbations of passing stars and interstellar gas clouds. The inner boundary, therefore, corresponds to the region where the greatest numbers of cometary nuclei are expected to reside. Proxima, being separated by 15,000 au from α Cen AB, sits slightly inside of the expected inner boundary of the Centaurian cometary cloud.

The time interval T_{imp} between successive Proxima impacts can be estimated from the number density ρ_{CC} of cometary nuclei in the Centaurian cloud, the relative velocity V_{Prox} of Proxima through the cloud, and σ the cross-section area, including a gravitational focusing effect, over which Proxima can sample cometary nuclei (see Appendix 2 in this book). Accordingly

$$T_{imp} = 1/\rho_{CC}\, V_{Prox} \pi\, R_{Prox}^2 \left[1 + V_{esc}^2 / V_{Prox}^2\right] \quad (3.1)$$

where $V_{esc} = 568$ km/s is the escape speed for Proxima, and R_{Prox} is Proxima's radius. Estimates for the relative velocity fall in the range of order 0.2 km/s. Substitution of characteristic numbers in Eq. 3.1 indicate that T_{imp} (years) $\approx 50/\rho_{CC}$.

The final step now requires an evaluation of the number density of cometary nuclei per cubic astronomical unit – and this unfortunately is not well known. Using our own Oort Cloud as a guide, however, we suppose that within the region from 10,000 to

[14]Hills assumes that the Oort Cloud-producing star systems fall in isolation. This may well be true for many systems, but the formation environment may additionally be very important. Not only might the size of an Oort Cloud region be restricted by the presence of nearby contemporaneously forming systems, but so, too, might the capture and loss of cometary nuclei from one system to another be an important factor in determining the overall cometary population.

20,000 au from the Sun there are of order 10^{13} cometary nuclei. Accordingly, the number density of cometary nuclei will be about 3 per cubic astronomical unit, and this suggests a cometary collision time interval of about 15 years for Proxima. Collisions occurring during this time interval will be very difficult to distinguish from ordinary flare activity on Proxima. If, however, the cometary nucleus density in the Centaurian system is as high as, say, 10 or 20 per cubic astronomical unit, then the time interval between collisions would be reduced to a more readily detectable 2.5–5 years.

The data on long-term trends in Proxima's activity variations are presently sparse, but Fritz Benedict (University of Texas at Austin) and co-workers[15] have used the Hubble Space Telescope fine guidance sensor to reveal an estimated 1,100 days (about 3 years) modulation period in its brightness. In contrast, an activity index period of about half that found by Benedict and co-workers, amounting to some 440 days (1.2 years), has been determined by Carolina Cincunegui (Insitituto de Astronoica, Argentina) et al. in a spectroscopic dataset collected between 1999 and 2006.[16] Cincunegui and co-workers specifically measured the characteristics of the ionized calcium (Ca II) H and K absorption lines in the spectra of Proxima to construct an activity index A – see Fig. 3.5. In the case of the Sun this index is taken to be a proxy measure of sunspot activity. It would generally be expected that variations in the activity index would be synchronized with those in brightness, as is the situation with the Sun, but the present data for Proxima does not support this conclusion. The full situation is currently unclear, and although it may be that Proxima's flare rate and activity level is modulated by cometary impacts, only the collection of more detailed observational data will tell what the true story is.[17]

Ultimately the best way to fully understand the small object, cometary and asteroid, populations within the Centaurian system

---

[15] Fritz Benedict et al., "Photometry of Proxima Centauri and Barnard's Star using the Hubble Space Telescope Fine Guidance Sensor 3: A search for periodic variations" (*Astrophysical Journal*, **116**, 429, 1998).

[16] Carolina Cincunegui et al., "A possible activity cycle in Proxima Centauri" (*Astronomy & Astrophysics*, **462**, 1107, 2007).

[17] Recall from Sect. 2.8 that the currently favored α^2 model for the generation of magnetic fields in stars like Proxima does not predict the occurrence of any specific periodic activity.

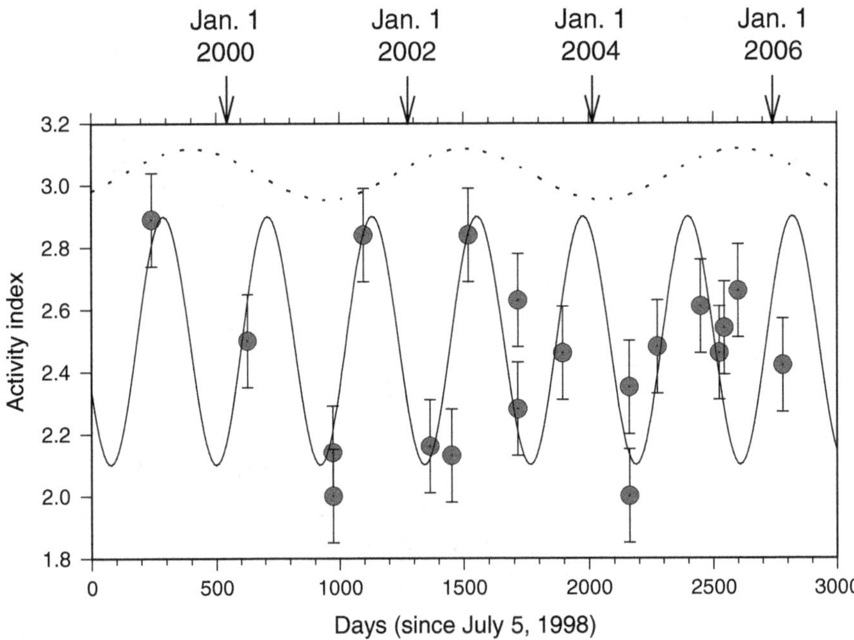

FIG. 3.5 The variation of the activity index as derived for Proxima Centauri by Cincunegui et al. (2007). The (*solid line*) sine curve shows the 422-day variation in the activity index for the data collected between March 1999 and February 2006. The upper (*dotted line*) sine curve shows, to an arbitrary scale, the 1,100-day brightness variations described by Benedict et al. (1998). (The time of minimum as observed by Benedict and co-workers is taken to be February 1, 1995)

will be to send interstellar spacecraft, both human-crewed and autonomous, to make in situ observations. This is clearly a long-term goal, and before it might be achieved the many problems associated with the organization and engineering of an interstellar space mission will need to be solved.

3.5 Getting There: The Imagined Way

We have seen earlier (recall Table 3.1) that over the past 80 years the sequential pens, typewriters and word processors of science fiction writers have produced a mixed set of predictions for the number of planets that might reside within the α Centauri system. It now seems appropriate to review, albeit but briefly, the means by which those same writers, as well as their companions from other literary and artistic genres, have enabled human beings to travel to the planets and star systems beyond our own. The underlying reason for presenting such a review is to see if there are any hints of practical methodologies by which space probes, even humans, might be cast free of the chains, so grimly described by van Vogt, that bind them to Earth.

Where to begin? The human journey of adventure, real and imagined, like a magnificent river, runs long and deep. And while, just like a river, the human imagination occasionally meanders, and slows, only to later crash headlong over turbulent rapids, or cascade down shimmering waterfalls, it always ends up in the sea – the vast ocean of human history and knowledge. For indeed, it has been estimated that the present human population, a staggering six billion people, carried upon Earth's now teetering shores, amounts to about 10 % of the entire complement of all human beings that have ever lived. The dead outnumber those alive today by ten to one, and the accumulated ocean swell of their thoughts, actions and wisdom offer rich pickings for the would-be angler of ideas. The collective consciousness of the now long dead, however, knew very little about the physical size of the heavens, nor indeed, its contents, but they knew how to dream, and their imaginations were no less potent than our own.

From an astronomical perspective, there seems no better starting place for our survey than with the works of Johannes Kepler. Indeed, Kepler was a great dreamer who dared to write on topics that other academicians, for a whole host of reasons, carefully avoided. The greatest of Kepler's imaginary works were written, on and off, over a time interval spanning some 41 years, and only saw publication in 1634, 4 years after his untimely death.

Kepler's *Somnium* was not only a book of dreams; it was a radical book of dreams concerning the Copernican perspective with both spin and orbital motion being attributed to Earth. The storyline, however, eventually requires the hero, Duracotus, to travel to the Moon, and to achieve this end a dream sequence is invoked, in which a demon guides the would-be adventurer across lunar space. The journey takes just four hours – implying, in modern terms, a speed of about 27 km/s. The mode of transport does not provide for a comfortable ride, with the traveler being subject to great cold, and to suffering a great initial acceleration, being "hurled just as though he had been shot aloft by gunpowder to sail over mountains and seas." For Kepler, therefore, the journey to the Moon is achieved via an occult-driven, out of body, dare one say teleportation-like mechanism.

Clearly, Kepler is not describing any *real* mechanism by which he believes space travel might be realized. Even a massive gunpowder-fired cannon, a device that he would certainly have been familiar with, is deemed insufficient to hurl a traveler skyward. Kepler well knew that something more powerful than the best percussive technology available in his time was required if space travel was to become a reality.

Jumping three centuries forward from the death of Kepler, the Christian apologist, philosopher of stentorian grace and multi-genre writer C. S. Lewis, in supposed competition with J. R. R. Tolkien, produced a series of three science fiction books that, among numerous settings and themes, required various characters to travel from Earth to Mars and then to Venus. Although Lewis was certainly not anti-science, he took no great pains to investigate possible means of interplanetary travel. In *Out of a Silent Planet* (published in 1938) we learn that the spaceship is simply powered by a 'subtle engine,' that is, by "exploiting the less observed properties of solar radiation." Lewis's statement, of course, means nothing specific, but in some sense it foreshadows the development of space sails (to be discussed later).

In Lewis's second book in the series, *Perelandra* (published in 1943), the transport becomes even more bizarre and occult, with the traveler being required to lie in a coffin-like container. The journey is described in terms of perceived colors and the sense of movement, although the observer describes it as gliding, "almost silently." Lewis was not mocking science in his choice of transportation, but rather he was attempting to provide a sense of wonder. Just as Tolkien was to emphasize in his *On Fairy-Stories* (first published in 1966, although initially presented as a lecture in 1939), so Lewis felt that it was through fantasy writing, and its linkage to the wonderful, that we can learn to re-interpret, and therefore see anew, the actual world in which we live. Fantasy literally strengthens and reinvigorates us, and for our present purposes, in terms of interstellar space travel, it is this mental invigoration that will prove crucial. With Lewis, and to a lesser extent with Tolkien, it is not engines and new physics that we find, but it is mind fuel and new vistas for our brains to interpret and our imaginations to shape.[18]

Moving back in time and away from occult forces to more natural ones, British historian and one-time Bishop of Hereford Francis Godwin invoked the use of tethered gansas (wild swans) to carry his imagined Spanish adventurer, Domingo Gonsales, to the Moon. Published in 1638, it appears that just like Kepler, and his long-term writing of the *Somnium*, so Godwin wrote, on and off, his *The Man in the Moone: or a Discourse of a Voyager Thither* over many decades. The craft that takes Gonsales to the Moon uses avian muscle power to travel, and the journey to the Moon was in some ways fortuitous and the result of the folklore belief that migratory birds actually wintered in the lunar realm. Ultimately, one might argue, that muscle power is really an expression of burning chemical energy, but, of course, it is now clear that flapping wings will be of no use for propulsion in the near perfect vacuum of space.

Not long after the appearance of Kepler's *Somnium* and Godwin's *The Man in the Moone*, the French playwright Cyrano de Bergerac saw into print his *The Other World: comical history of the states and empires of the Moon* (published in 1656). This satirical first person narration supposedly concerned de Bergerac's 'actual' adventures on the Moon, and it addressed the issue of space travel via experimentation. First of all the would-be adventurer straps jars containing morning dew to his belt – the idea being that as the dew evaporates so his body will be lifted skyward. This mechanism only succeeds in transporting the traveler to what is now Quebec (then New France) in Canada. A second attempt to reach the Moon is made with a winged flying machine, but this too initially fails – "I fell with a sosh in the valley below," writes de Bergerac. Eventually, however, the flying machine is successfully launched skyward by strapping numerous fireworks to its frame. The combined thrust of the fireworks turns out to be

[18]Martha Sammons explores in detail the idea of fantasy driving the imagination, leading to re-evaluation of the real world, in her work *A Guide through C. S. Lewis' Space Trilogy*, Cornerstone Books, Westchester, Illinois (1980).

sufficient to launch de Bergerac on his way, although the machine itself eventually falls back to Earth. With de Bergerac we are slowly moving towards the idea of a space rocket in which the propulsion is generated via a chemical reaction (i.e., the rapid burning of gunpowder).

French author Jules Verne in his 1865 novel *From the Earth to the Moon* reverts to massive cannon power to launch his explorers into space. To this end, however, an extraordinarily large columbiad is required to do the job, and the Moon capsule is fired from a 900-ft-deep pit filled with 400,000 lb of gun cotton. Given that the launching mechanism is located on Earth, the narrative developed by Verne is that of a one-way trip, and the story ends with the hapless explorers being trapped in eternal orbit around the Moon.[19] Although Verne valiantly attempts to provide a detailed mathematical and physical account of the cannon's characteristics, so that it might successfully launch the space capsule to the required escape speed from Earth, he completely fails (in realistic terms) to account for the survival of the passengers during takeoff – for, indeed, "the firing of the cannon was accompanied by a veritable earthquake. Florida was shaken to its entrails."

Less technically descriptive than Verne, Herbert G. Wells, in his classic invasion-story *The War of the Worlds*, published in 1898, also adopted ballistics over rocketry for the means by which the Martians came to Earth: "An enormous hole had been made by the impact of the projectile... the cylinder was artificial – hollow – with an end that screwed out...." For the Martians it was a one way trip to invasion and war. French film director George Méliès in his pioneering *Le Voyage dans la Lune* (released in 1902) also struggled with the re-launch problem of his astronaut carrying giant shell of a spacecraft from the Moon; its homeward plunge being initiated by the physically unrealistic method of falling off a lunar mountain.

Writing 15 years after Verne, British author Percy Greg introduced a totally new idea for powering a spacecraft in his 1880

[19] Verne eventually produced a sequel to *From the Earth to the Moon,* in the form of his (1870) publication *Round the Moon.* In grand literary style Verne completely ignores the predicament that he had placed his hapless travelers in at the close of his earlier novel, engineering the storyline to bring about their eventual safe return to Earth.

novel *Across the Zodiac: the Story of a Wrecked Record*. Not only is the propulsion idea new, Greg also envisions a voyage to the planet Mars,[20] since the Moon, according to the narrator is, "a far less interesting body." The spacecraft, somewhat oddly called Astronaut, is powered by a substance called apergy, which Greg explains is a "repulsive force in the atomic sphere" that can be collected, stored and discharged through "the progress of electrical science." There are, of course, two nuclear forces that operate within atomic nuclei[21] – not that physicists knew of them in the 1880s – but how any such 'forces' might be extracted and utilized directly in the form of propulsion is a complete mystery. This being said, it could be argued, although perhaps not in any spirited fashion, that Greg foresaw the development of the fusion drive (to be described below) that is dependent upon the interaction of atomic nuclei. Greg's imagination with respect to spacecraft propulsion, although very vague on detail, is nonetheless all the more remarkable, given that when he wrote his story the invention of the steam turbine (by Charles Parson, in 1884) was still 4 years' distant, and Henry Ford was a further 28 years away from introducing his Model T automobile.

Following in Greg's footsteps, H. G. Wells in his short novel *First Men in the Moon*, published in 1901, additionally takes a wholly new tack in spacecraft design and propulsion by invoking the serendipitous development, by Mr. Cavor, of a gravity-shielding paint (appropriately called cavorite) on "14 October, 1899."[22]

[20] Greg appears to be the first science-fiction author to include mention of the newly discovered moons of Mars. Writing some 2 years after their first detection by Asaph Hall, Greg writes in the voice of the unnamed narrator, that observing from the spaceship Astronaut, "I discovered two small discs, one each side of the planet.... evidently very much smaller than any satellite with which astronomers are acquainted... they were evidently very minute, whether 10, 20 or 50 miles in diameter I could not say". See also Note 6 above.

[21] The two nuclear forces are the strong force, which holds the nucleus of protons and neutrons together, and the weak force, which enables atomic decay and nuclear fusion.

[22] The date that Wells provides for the invention of cavorite is very precise, but I have not been able to find any specific reason for it. The year in question was certainly a dramatic one for Wells in that during its course he suffered several nervous breakdowns. Towards the close of 1899, however, due to better than expected book sales, he was able to sign a contract to begin the construction of a custom-built house, Spade House, in his home county of Kent. I tentatively suggest that the money derived from and the newfound success of his writings, lifted a great metaphorical weight from

This idea is something altogether new, and it moves the propulsion concept away from those employing chemical reactions, or other reactive agents, which, for an Earth launch, must work against the force gravity.

Friedrich Mader also invokes a form of anti-gravity drive for his spherical spaceship, called Sannah, in his 1932 novel *Distant Worlds*. Mader's drive is somewhat different from the cavorite invoked by Wells, in that it uses a "combination [of] electrical, or a magnetic current" to generate a "centrifugal power" that acts in opposition to gravity. It is not a gravity shield as such but a motive force that causes objects to move apart. Isaac Asimov in the last of his Foundation novels, *Foundation and Earth*, published in 1986, also invokes the idea of gravity shielding as a means of allowing spaceflight. (He also allows, via unspecified means, for so-called hyperspatial travel between stars.) Recognizing the physical difficulties attached to any kind of gravity-shield drive, Asimov, out of necessity, hides behind literary camouflage and simply notes that the spacecraft, the Far Star, has the "capacity to insulate itself from outside gravitational fields to any degree up to total.... the gravitational effect *within* the ship, paradoxically remained normal." In some sense Asimov appears to be suggesting the drive works by constructing a moveable gravity bubble; the gravitational field within the bubble being non-zero and constant with the gravitational field at its outer surface being zero.[23] The problem, of course, is how might something like cavorite, centrifugal power or a gravitic drive be developed? The answer to this question is simply that we have absolutely no idea, and nor, sadly, did Wells, Mader or Asimov. For indeed, gravity may well be the weakest of the known fundamental forces of nature, but it is all

---

Wells' shoulders and that indeed it was towards the close of 1899 that he may have felt that he was free to fly literarily, unfettered by the gravitas of money problems. The Wells family formally moved into Spade House in 1901, and it was from there that he produced many of his most famous works.

[23] Asimov's idea is actually the exact converse of a classical result revealing that the gravitational force on a point mass located inside of a thin spherical shell is exactly zero. This result is encapsulated within Newton's Shell Theorem, as proved by Isaac Newton in his 1687 *Principia*. The key point of the theory is that it allows the important simplification that the entire gravitational mass of an extended, spherically symmetric object (e.g., a planet or a star) can be thought of as a point mass located at the body's center.

pervasive, acting over all distances and upon anything that has a mass – that is, anything made of atoms or stable elementary particles.

As we move deeper into the first quarter of the twentieth century, aircraft technology, rocket engineering and the appeal of space travel advanced considerably. The Wright brothers realized the first controlled and sustained flight of a heavier than air aircraft in 1903, and such space pioneers as Russian mathematician Konstantin Tsiolkovsky and American physicist Robert Goddard were in their imaginative and experimental ascendancy, planning possible space missions, conceiving multi-stage rockets and building new engine configurations.

These same researchers began the process of moving away from the methodology of gunpowder-fired rockets to those using liquid fuels, and to the development of rocket engines fitted with carefully designed and sculpted exhaust gas constraining and accelerating De Laval nozzles.[24] Such new developments not only inspired the work and imaginations of numerous scientists and engineers, they also oversaw the beginnings of the golden age of science fiction writing, inspiring authors and readers alike to dream of space adventure, new worlds and first contact with alien civilizations. The first issue of *Amazing Stories* appeared in 1926, and in 1929 it was followed by *Science Wonder Stories*. These early magazines specialized in publishing short stories of the weird and wonderful, as well as the mad and mind bending that might reside in the vastness of the universe. These same stories opened up to the imagination the prospects of imminent interstellar space travel, and inevitably science fiction began a slow and steady transformation into science fact.

In 1927 the German Spaceflight Society (Verein für Raumschiffart) held its inaugural meetings, and a young Wernher von Braun, later the great driving force behind the development of the Saturn V rockets that took the first humans to the Moon, was to join its ranks in 1930. Likewise, the renowned British Interplanetary

[24] The De Laval nozzle has a carefully engineered hourglass profile that results in enhanced pressure and acceleration, to supersonic speeds, of any hot gas flowing through it. The nozzle was developed in the late nineteenth century by Swedish engineer Gustaf de Laval. Not only functional, it has also become the iconic minds-eye image of what a rocket engine exhaust nozzle should look like.

Society was formed in 1933, and across the Atlantic, just a few years earlier (in 1930), an enthused group of science fiction writers had established the American Interplanetary Society (re-named the American Rocket Society in 1934). Indeed, the 1930s was a time of great scientific and engineering innovation, and it was a time to dream of space adventure. It is within this early timeframe that we find the publication of such naively wonderful works as *By Rocket to the Moon* (published in 1931) by German journalist Otto Gail – a story not only involving high adventure but also making use of and describing new concepts in aircraft and rocket design. Gail specifically describes the pioneering designs then being developed by Austrian engineer Max Valier.

By the mid-1930s an incredible degree of foresight and innovation had taken place, and one finds, for example, Murray Leinster describing in his 1935 short story "Proxima Centauri" a vast spaceship, a world in its own right, which houses a self-sustaining ecosystem, a living habitat for its many passengers (which is environmentally controlled) and a fleet of small robotic spacecraft to carry out any required external repairs. Furthermore, Leinster also describes a spaceship that can make the journey to Proxima Centauri in just 7 years, thereby implying the attainment of a speed equal to about half that of light. Anticipating the criticism that not even the most optimistic projections for rocket engine development could propel a spaceship to such a high, relativistic speed, Leinster invokes a new form of force field drive powered by the "disintegration effect of the Caldwell field"[25] – a scientific copout, yes, but a literary necessity.

Nearly a quarter-century on from Leinster, Edmund Cooper, in similar vague manner, describes the construction of a massive

[25] Leinster provides in fact a great deal of information about his Caldwell field drive. "At full acceleration" he writes, the drive "disrupted 5 cubic centimeters of water per second" and this disruption "collapses electrons of hydrogen so that it rises in atomic weight to helium, and the helium to lithium, while the oxygen of the water is split literally into neutronium and pure force." This all sounds technically wonderful, but it is sadly a literary sham, and has absolutely no connection with real-world physics. It is possible that Leinster named his force-field drive after Eugenie Caldwell (1870–1918), who was an American electrical engineer, later medical doctor, who pioneered the development of medical X-ray radiology. Wilhelm Roentgen's discovery of X-ray radiation in 1895 and its ability to reveal images of bones under a covering of skin and sinew was a marvel of the times. Often described as a martyr of science, Caldwell died of radiation-induced skin cancer in 1918.

interstellar spaceship in his reflective but dystopian novel *Seed of Light* (published in 1959). It was a "self-contained world which might be required to support human life independently for centuries... powered by volatility rockets and sub-atomic motors." It took the ship, the Solarian, 30 years to journey to α Centauri, indicating a speed just in excess of a tenth the speed of light c (with $c = 3 \times 10^8$ m/s).

Less speedy than Cooper's Solarian, but no less spectacular in scale, is the Centauri Princess, envisioned by A. Ahad in his *First Arc to Alpha Centauri* (published in 2005). This spacecraft, modeled after the O'Neill cylinder concept (as developed in the mid-1970s), is described as being some 9 miles long and 6 miles in diameter, and travels at a stately 60,000 mph (27 km/s = 10^{-4} c). Driven by "nuclear powered engines," working along the lines suggested for the Orion Project atomic spaceship (described below), the Centauri Princess is described as being a self-contained, artificial world with a crew of 3,000 people. In addition, the crew/population is described as being a multi-generational one, since the total travel time to α Centauri is envisioned as being of order 40,000 years.

Albert Einstein introduced his ideas of special relativity and general relativity in 1905 and 1915, respectively, and although they revolutionized physics they effectively terminated the dream that humans (or indeed any sentient beings) might freely travel among the stars. Limited, both practically and physically, to speeds much less than that of light means that stellar travel times are not just long, they are multi-generational. For the solar neighborhood (recall Fig. 1.18), and for α Centauri specifically, this speed limitation is perhaps not a fatal issue, since even at a tenth the speed of light, a realistically achievable speed in the modern era, at least a one-way trip might be realized in half a human lifetime. Just as we live with the expectation of innovation and new developments in the modern era, so the science fiction writers of the mid-twentieth century played with the idea of technology jumps – literally near-instantaneous leaps in engineering and mechanical ability.

E. van Vogt, for example, explored this very idea in his short-story *Far Centaurus* (first published in 1944). It was, Vogt writes, the development of the "eternity drug" that opened up the possibility of space travel. By inducing an ageless sleep, human explorers could be sent on interstellar space missions by conventional, that is, relatively slow, rocket-powered means. Accordingly, a ship leaves Earth, with a sleeping cargo of four astronauts, on a 500-year journey to α Centauri. Unfortunately for the deep-dreaming astronauts, just a century and half after they leave Earth, an interstellar flight drive is invented, which cuts the α Centauri travel time to just three hours! When the astronauts are roused upon their arrival at α Centauri, not only have four planets there been named after them, but these same planets support vast colonies of human beings.

Indeed, in his narrative van Vogt essentially foresaw what has more recently become known as the incessant obsolescence postulate. Under this postulate it is argued that there is no point in starting any interstellar space mission 'now' since future advances in technology will inevitably make for faster travel speeds and shorter travel times. The problem with buying into this postulate *holus bolus*, however, is that one never actually does anything.[26] Indeed, once entwined within such a mindset, it is both physically and psychologically difficult to break free. The logic of the postulate is compelling, but it simply results in moribund inactivity. Perhaps the only reasonable way to escape the chains of the obsolescence postulate is to forge at some specific epoch a minimum series of realistic and achievable thresholds that, upon being breached, clearly establishes the result that the time, that all important 'now,' for the initiation of interstellar space travel and exploration has arrived. As we shall discuss more fully below, many present-day researchers believe that the epoch-defining 'now' is virtually upon us, and that (un-manned) interstellar space missions will be initiated within the timeframe of the next century.

[26] Perhaps the ultimate example, albeit in a literary form, in which the adoption of the incessant obsolescence postulate proved successful is that of the infinite improbability drive as envisioned by Douglas Adams in *The Hitchhikers Guide to the Galaxy* (published in 1979). In this case a fully functional, faster than light spaceship drive simply appeared one day when it was realized that it had a finite rather than an infinite improbability of existing.

The staggeringly short three-hour travel time to α Centauri invoked by van Vogt in *Far Centaurus* is an impressive transgression of special relativity dictates, since the implied spacecraft speed is nearly 13,000 times faster than that of light. This being said, they are technically not in absolute violation of general relativity. General relativity is Einstein's description of how spacetime, the four-dimensional space plus time coordinate system within which events can be located and described, is structured. Specifically, it describes how the geometry (or shape, if one likes) of spacetime is altered by the distribution of mass and energy. A massive object, for example, causes spacetime to curve, and this curvature is made manifest through the accelerated motion of nearby smaller mass objects. Isaac Newton interpreted accelerated motion in terms of a gravitational force; Einstein, in contrast, argued that the gravitational force is really an illusion, an illusion that comes about because of objects moving within the curved geometry of spacetime.

Be this as it may, what is more useful to the imagination, it turns out, is that distinct regions of spacetime, regions, say, on opposite sides of the galaxy that would take even light tens of thousands of years to cross, can in principle be connected by a shortcut bridge, or as they are more commonly called a wormhole. The shortcut pathways, again in principle, essentially enable one to move between distinct regions that might be vast distances apart, almost instantaneously.

An alternative to the wormhole, shortcut spatial bridge, has been described by author Rick Novy in his fictional work *Rigel Kentaurus* (published in 2012). Here the "Mudge drive" is described, as a device that once activated, "tears open a hole in space and time," allowing the spacecraft to "travel in the direction of the Big Bang, when the universe was smaller. With a smaller universe, everything is closer together, and the drone [a faster-than-light spacecraft] can travel great distances in significantly less time."

Clearly, in terms of interstellar, even intergalactic, travel, the ability to manipulate spacetime would be highly useful. The problem, of course, is that no such skills or objects such as macroscopic wormholes really exist, or will ever likely exist.

They are just mathematical, albeit highly complex mathematical, daydreams.[27] Perhaps the author is allowing a failure-of-the-imagination moment to creep in with the latter sentence, but it does seem that to suggest the use of such hyper-unrealistic structures, within the context of space travel and exploration, is at best wishful thinking and at worst detrimental to the present-day call for the initiation of deep space missions. Likewise ideas invoking the application of warp drives, where spacetime is physically manipulated so that an object might travel at speeds greater than that of light, should also be discarded – mathematically demonstrable on paper within the context of the present theory general relativity,[28] yes; a playful piece of mathematical wizardry to make us smile, yes; a very useful device for advancing the storyline of a *Star Trek* or *Star Wars* film, yes; but a physically realizable solution to interstellar space travel, no. Warp drives, gravity shields, hyperspace and wormholes offer nothing of utilitarian substance to the current, even the foreseeable far future debate on interstellar space travel; they give the pretence of somehow enabling cosmic adventure, but in reality they simply provide false hope, leaving us free-wheeling on a stationary bicycle.

One form of advanced propulsion drive that may yet prove its mettle in foreseeable centuries is that in which energy is generated through matter-antimatter annihilation. This very process is at play, in fact, in all main sequence stars generating their energy through the proton-proton chain and CN cycle conversion of hydrogen into helium (recall Figs. 2.3 and 2.4). Specifically it is the fate of the positron e^+ to annihilate with its antimatter particle, the electron e^-. The annihilation provides energy in the form of

---

[27] American physicist John Wheeler (1911–2008) is one of the most respected scientists of the entire twentieth century, and his many writings and ideas are always worthy of attention. In a biographical account, however, Wheeler made the extraordinary claim that one of the basic working assumptions adopted throughout his scientific career was that nature will always find a way, sooner or later, of exploiting every feature of any correct and allowed physical theory: "If relativity is correct," he writes, "and if it allows for wormholes, then somewhere, somehow, wormholes must exist – or so I want to believe." These are certainly strong statements, not to be taken lightly, but the key terms in Wheeler's statement are the 'ifs,' and it is presently far from clear if general relativity is the correct theory to apply at the level where wormhole formation might be allowed.

[28] M. Alcubierre, "The warp drive: hyper-fast travel within general relativity" (*Classical and Quantum Gravity*, **11**, L73, 1994).

two gamma ray photons: $e^+ + e^- \rightarrow \gamma + \gamma$, with the energy E being carried away by the two gamma rays being equal to $E = 2m_e c^2$, where $m_e = 9.109 \times 10^{-31}$ kg is the mass of the electron.

All atomic particles have an antiparticle companion, and their meetings always generate an explosive outburst of energy. The problem, and of course there is always a problem with advanced propulsion mechanisms, is that we live in a matter-dominated universe. Positrons are both generated and then rapidly destroyed within stars, and problematically there is no free source of antiparticles (not just within the Solar System but anywhere in the universe) that can be simply 'mined.' Antimatter must be generated (a complex process requiring large amounts of input energy) and then stored – and stored very carefully, since even the slightest leak in the confinement container will result in the explosive destruction of the container.

The idea of utilizing an antimatter-matter annihilation engine to power an interstellar spacecraft has been around for many decades. One such example is that invoked by Charles Pellegrino in his 1993 novel *Flying to Valhalla*. The Valkyrie spacecraft is, in fact, a concept designed and developed by Pellegrino and James Powell (then working at the Brookhaven National Laboratory – and perhaps more widely known today for the 1968 invention, with Gordon Dandy, of the maglev train) in the mid-1980s. Powered primarily by proton-antiproton annihilation the Valkyrie spacecraft will (theoretically) achieve speeds close to 90 % the speed of light, making a trip to α Centauri a mere 4.5-year cruise. Allowing for an acceleration phase and spacecraft deceleration upon arrival, Pellegrino suggests a one-way journey to α Centauri might take about 7 years to complete. In *Flying to Valhalla* Pellegrino predicted that field testing of the antimatter drive for the Valkyrie spacecraft would begin in 2008, and that the first crewed spacecraft would leave for α Centauri in 2048. His present Internet webpage indicates that "The Valkyries should be flying by the year 2070." These are indeed bold dreams, but a long way yet from being a financed or practical reality.

Is the current obsession with maximizing travel speed and the concomitant development of new and highly sophisticated technologies really the best way to go about interstellar space travel? Perhaps, a slower and gentler approach is better and more realizable. To this end solar sails and vast clipper-ship designs have been touted as one way, with near-contemporary technology, that the journey to interstellar space, or at least deep Solar System space, might be obtained. The idea of sailing into space, by design or accident, has an ancient heritage, and satirist Lucian of Samosata (A.D. 125–180) described one such adventure in his *True History*.

This ill-titled work concerns the trials and tribulations suffered by a company of seagoing explorers who, on one occasion, become caught up in a sudden whirlwind. After being lifted 3,000 furlongs into the air (some 603.5 km by modern measure), the adventures, after 8 days of buffeting, are conveyed to the Moon, whereupon they are caught up in a tumultuous battle raging between the king of the Moon and the king of the Sun. A similar journey to Lucian's adventurers, although this time taking 6 weeks to accomplish, is described in *The Surprising Adventures of Baron Munchausen*, published between 1781 and 1783. Indeed, in the entirely (un)believable adventure the good Baron explains, "I went on a voyage of discovery at the request of a distant relation, who had a strange notion that there were people to be found equal in magnitude to those described by Gulliver in the empire of BROBDIGNAG." And, of course, Johnathan Swift's *Travels into Several Remote Locations in the World* has its flying island of Laputa – an island held aloft by magnetic levitation.

Unlike the heavy canvas sails that carried Lucian's adventurers and Baron Munchausen to the Moon, modern solar sails are microscopically thin and extremely lightweight, and, of course, they gain their motive force through interacting with the Sun's radiation field – or potentially through interacting with a powerful beam of microwaves. Indeed, it is the momentum transfer of reflected light that powers these celestial ships. Many space sail designs have been proposed over the years, but one specific design, the Starwisp, by American physicist, aerospace engineer and science fiction writer Robert Forward, has a specific elegance worthy of detailed study (Fig. 3.6). The Starwisp was designed not only to undertake a trip into interstellar space but also for a flyby mission to α Centauri. How such a mission might unfold is illustratively described by Stephen Baxter in his 2013 book *Proxima*.

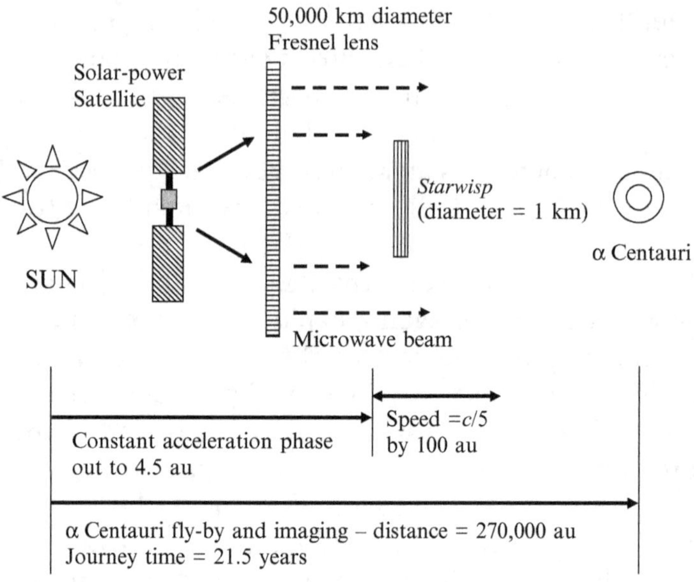

FIG. 3.6 Robert Forward's Starwisp interstellar space-sail mission concept (R. L. Forward, "Starwisp – An ultra-light interstellar probe" (*Journal of Spacecraft and Rockets*, **22**, 345, 2985))

The Starwisp project began as the result of a chance encounter and subsequent discussion between Forward and Freeman Dyson (who we shall encounter again). The topic of discussion focused on interstellar transport, and a technical question related to the idea of making space sails lighter by cutting holes smaller than the wavelength of light into their fabric. According to Forward,[29] "Dyson produced some notes from his files on an interstellar perforated sail pushed by microwaves." It was these notes that inspired the Starwisp design. Indeed, rather than being made of a single mono-film substrate, the Starwisp sail is really a wire mesh, the wire strands crisscrossing like the warp and weft in a loom. The spacing of the wires in the sail mesh are laid out so that they will specifically interact with microwave radiation with a wavelength of 3 cm. In this situation, it is not so much starlight that will drive the sail but rather microwaves beamed from a solar-powered satellite that will be used as the accelerating agent.

[29] R. L. Forward, "Starwisp – An ultra-light interstellar probe" (*Journal of Spacecraft and Rockets*, **22**, 345, 2985).

Indeed, Forward additionally invokes in his research paper the construction of a 10 GW microwave transmitter, located aboard a Sun-orbiting satellite, that with the aid of a 50,000 km Fresnel lens[30] will set the Starwisp on its journey to α Centauri. The Starwisp sail will be just 1 km across and weigh in at just 20 g! The characteristic acceleration of such a lightweight sail, if driven by a 10 GW microwave beam, is remarkably high and within about a week it would be traveling at about a fifth the speed of light. At this cruising speed the journey time to α Centauri is 21.5 years. Not only does the wire mesh sail provide for the motive force for the Starwisp, it also acts as the power conductor and the connecting network for the sail's microcircuits (transmitter, cameras, science packages and guidance system).

The Starwisp is a wonderfully complete, integrated and innovative design, and sadly, while Forward comments in his 1985 research paper that, "If we desired, the first Starwisp probe could be sent to Alpha Centauri before the millennium is out," no such mission was, nor has since been, adopted. The Achilles heel in Forward's design, as far as seeing a mission launch, is not the building of the Starwisp itself, but rather the construction of the power satellite to provide the accelerating force. We are no nearer now, in 2014, to seeing a working solar-driven microwave power satellite being put into operation than we were in 1985.

Although Forward's Starwisp program is now some 30 years behind schedule, Lou Friedman, co-founder of the Planetary Society, does not provided us with any great hope that a space sail mission will be launched anytime soon. Indeed, Friedman argued in 2007 that,[31] "Practical interstellar space flight [via light sails] is at least 2 centuries in the future." Time, as ever, will tell how all this plays out. It seems appropriate, however, to remind ourselves of the words of Baron Muchausen, "I know these things appear strange; but if the shadow of a doubt can remain on any person's mind, I say, let him take a voyage there himself, and then he will know I am a traveller of veracity."

[30] Developed by French physicist Augustin-Jean Fresnel (1788–1827), a Fresnel lens typically has a large aperture and short focal length. The lens was originally developed for focusing and projecting the light beam that was to be emitted from a lighthouse.

[31] See Louis Friedman's article, "Making light work" (*Professional Pilot magazine*, June issue, 2007).

By the close of the 1950s all of the basic possibilities by which interstellar space travel might be initiated had been imagined – from chemical rockets, along with gravity shields, to anti-gravity drives, force fields and riding the shock waves of sequentially exploded nuclear bombs. All, on paper at least, could (or might at some date) do the job of lifting and powering a spacecraft, the only issues undetermined were speed and mission longevity, along with the more practical problem of who was going to pay for it. With the advent of the Apollo missions to the Moon in the 1960s and 1970s, enthusiasm and bravado were in their ascendancy, and numerous space missions to Mars and the outer planets, as well as to the stars beyond, were developed. The timetables were ambitious, and it was envisioned that permanently manned bases would be in place on the Moon as well as Mars by the beginning of the twenty-first century. Evidently, we are living in the wrong future, and for political as well as financial reasons the dreams of yesterday's mission planners were never funded and eventually all came to naught. If humanity fails to construct and launch interstellar spacecraft it will not be through a lack of imagination and/or engineering skill.

3.6 And the Zwicky Way Is?

In terms of sheer audacity and far-reaching brilliance, the ideas of Swiss-American astronomer Fritz Zwicky (1898–1974) are always worthy of scrutiny.[32] A self-confident, original, often abrasive and self-described "lone-wolf" researcher, Zwicky championed the idea of discovery through what he called a morphological approach. This method might broadly be described as taking an holistic view to problem solving, its methodology being to explore "the totality of all of the possible aspects and solutions of any given problem."[33]

Beginning in the late 1940s and throughout the 1950s, Zwicky began promoting what might be called a grand assembly model for

[32] See, for example, Keith Cooper's biographical account, "Astronomy's Lone Wolf" (*Astronomy Now* magazine, February issue, 2014), and Stephen Mauer's "Idea Man" (Beamline magazine, Winter 2001; the article is available at www.slac.stanford.edu/pubs/beamline/31/1/31-1-mauer.pdf.)

[33] Fritz Zwicky, *Morphological Astronomy* (Springer-Verlag, 1957).

reaching the stars. Having spent many years working on rocket engine design, and upon the formulation of rocket engine fuel, Zwicky presumably came to the conclusion that interstellar travel via spaceships was neither practical nor feasible: "For the purpose of traveling to the nearest stars, Alpha Centauri for instance, at a distance of 4 light years, rockets do not suffice," he wrote in 1969. Undaunted by such mechanical restrictions, however, Zwicky then suggested that perhaps the best way to study interstellar space would be to move the Sun and the entire Solar System to the stars directly. In this way, we (that is, humanity) remain safely ensconced on Earth, with its habitable, Sun-heated surface and atmosphere intact, and simply carry on our everyday business as the journey proceeds. Zwicky specifically noted that, "Traveling at a speed of 500 km/s through space, relative to the surrounding stars, we might reach the neighborhood of Alpha Centauri in about 2,500 years."

Although this approximate travel time, given the speed stated, is correct, Zwicky (sadly) is somewhat vague on exactly how the Solar System might be accelerated to such a speed, although he essentially argues that the Sun itself might be turned into the engine. Zwicky's idea was to fire "solid particle pellets with velocities up to 1,000 km/s" into the Sun's atmosphere, to thereby ignite a localized region of surface nuclear fusion. The fusion reactions would then "eject [matter] with velocities of the order 50,000 km/s, while the resulting force of reaction would propel the Sun in the opposite direction." The idea is remarkable in its inherent simplicity, one could say, a mere application of Newton's laws. The idea is also remarkable for its inherently unlikely emergence; indeed, it is a triumphant result of Zwicky's morphological methodology. The practical problem, of course, is how do you construct and fire the ultra-high-speed pellets, constrain the surface fusion region and channel the eject material in the right direction on a repeated basis and in the right direction so as to achieve the required acceleration? The idea is bold, brash and, of course, completely impractical, but breathtaking in its overall outlook.

For all of its missing solar-engine detail, Zwicky's basic idea for interstellar exploration is far from being a stillborn vision. Indeed, it resonates with several more recently suggested ideas for both engineering the Sun, to substantially prolong its main-sequence lifetime, to terraforming the planets and for making moons and asteroids within the Solar System habitable. British astrophysicist Martyn Fogg, for example, described in the late 1980s a star-lifting mechanism based upon the manipulation of a star's mass-loss rate, period of rotation and magnetic field.[34] Viroel Badescu (University of Bucharest) and American geographer and macro-engineering specialist Richard B. Cathcart, at the turn of the last millennium, further studied the possible designs of stellar engines that might be employed by Kardashev Type II civilizations.[35]

The concept of solar sailing was perhaps taken to its limits, in 1987, by Russian physicist Leonid Shkadov, who investigated the possibility of using a truly massive mirror, or light sail statite, to guide and control the motion of the Sun, along with the attendant Solar System, across the galaxy – echoing Zwicky's grand vision. Shkadov's idea would take macro-engineering to its ultimate limits, and he envisioned the construct as a massive mirror at a stand-off distance of 3 au from the Sun. Thus positioned the solar radiation intercepted and subsequently reflected by the mirror would produce a small but constantly acting net force that would perturb the Solar System's trajectory around the galactic center (Fig. 3.7). The mirror would be designed so that its stand-off distance from the Sun remained fixed, hovering as it were like some ghostly spectra, held in balance between Sun's gravitational force and the force due to the Sun's radiation pressure. In terms of mirror properties, Shkadov concluded that it would require a minimum mass of order 10^{19} kg and a surface density of 1.55×10^{-3} kg/m^2. These requirements translate into what is ostensibly the construction of a mirror with sides 80,000,000 km long – a nontrivial engineering exercise indeed, requiring in essence the

[34] Martyn Fogg, "Solar exchange as a means of ensuring the long-term habitability of Earth" (*Speculations in Science and Technology*, **12**, 153, 1988). Details are also provided by the author in, *Rejuvenating the Sun and Avoiding Other Global Catastrophes* (Springer New York, 2008).

[35] V. Badescu and R. B. Cathcart, "Stellar engines for Kardashev's Type II civilizations" (*Journal of the British Interplanetary Society*, **53**, 297, 2000). Russian astronomer Nickolai Kardashev introduced the idea of a technology type in 1964. Three civilization numbers were introduced with labels I, II and III being assigned according to the control of materials and energy resources at the level of planetary, host star and host galaxy, respectively.

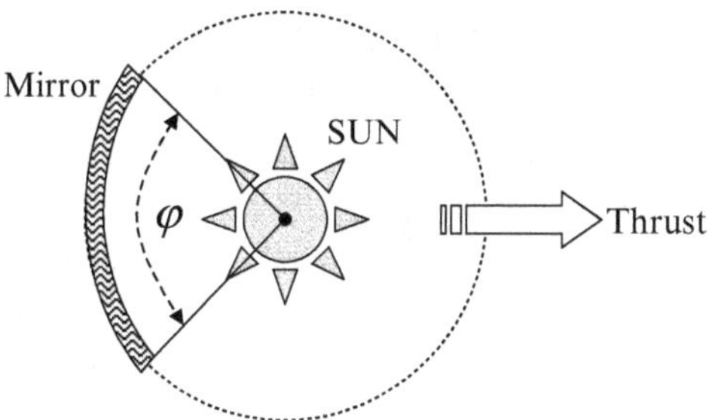

FIG. 3.7 The Shkadov thruster, class A stellar engine. The thruster is a spherical mirror arc spanning an angle φ. Radiation reflected by the mirror produces a radiation pressure imbalance that results in a net thrust (shown by the solid arrow). The force imbalance produced by the mirror will be of magnitude $F = (L_{\odot}/2c)\,[1 - \cos(\varphi/2)]$, where $L_{\odot}$ is the Sun's luminosity and c is the speed of light

controlled manipulation of the dispersed matter content of a 190-km rocky asteroid.

Viroel Badescu and Richard Cathcart further refined Shkadov's analysis in a set of detailed calculations published in 2006. These calculations considered the perturbed motion of the Sun within the galaxy's gravitational field, and indicated that a Shkadov thruster (or similar such stellar engine) could potentially shift the Sun and Solar System some 10 pc from their otherwise unperturbed position in a time span of order 150 million years. The possible construction of a Shkadov thruster, or any other Sun manipulation engine, is set well into our distant future. But, one can ask, might a more advanced civilizations have utilized the same idea? This very possibility, remarkably, is testable with current (and near future) survey data relating to the observations of planetary transits.

To this end, Duncan Forgan (University of Edinburgh) has recently studied the possibility of detecting extraterrestrial stellar engines via their masking effects.[36] Essentially, if a mirror partially covers the disk of a star in our line of sight, then the light curve

[36] See, Duncan Forgan (University of Edinburgh), arxiv.org/abs/1306.1672. This particular work builds upon the theme of Dysonian SETI – named after Freeman Dyson – in which it is argued that evolved technological civilizations will eventually begin to construct megastructures of one form or another within their natal planetary systems.

observed during a transit would likely be asymmetrical. Forgan concludes, that while there is no current evidence to suggest that any of the known transiting exoplanetary systems contain large Shkadov thruster-like structures, being aware of the possibility that they might establishes a potentially new way of looking for extraterrestrial intelligence. Indeed, it is by observing the light reflected from a 20,000 km light sail that astronomers on Earth first discovered the presence of the Phelan scout-craft in Michael McCollum's *The Sails of Tau Ceti* (Ballentine Books, New York, 1992). McCollum also explores the possible consequences and actions of two alien civilizations, one much more advanced than the other, making first contact.

3.7 It Will Not Be We...

"It will not be we who reach Alpha Centauri and the other nearby stars. It will be a species very like us, but having more of our strengths and fewer of our weaknesses." So wrote American astronomer Carl Sagan in his remarkable book, *The Pale Blue Dot: A Vision of the Human Future in Space* (Random House, New York, 1994). Here, of course, Sagan is really drawing attention to the social, economic and political issues that must first be resolved before the advent of interstellar exploration can really begin. Science fiction may well be our muse, and possibly even our guide, but the journey to the stars, even the very closest one after the Sun, will only be achieved by solving extremely complex engineering and design issues. In the interstellar travel realm, ideas are in fact cheap and numerous, while practical solutions are highly expensive and extremely limited.

There are essentially two issues that have to be dealt with when it comes to space propulsion. First, the spacecraft (at least historically and continuing into the present and near future) has to get from Earth's surface into some form of low-Earth orbit. In this sense, the spacecraft has to overcome the pull of Earth's gravitational field. This is the big struggle stage. Second, once in low-Earth orbit, even with the benefit of the inverse square law decrease in the gravitational force, the spacecraft has to initiate its journey into deeper space. Now the speed must be increased to above

Earth's escape velocity, and for interstellar space beyond the Solar System's escape velocity.[37] The speed that a spacecraft can achieve is determined (among numerous factors) by its physical mass and on the type of engine that it has. Clearly, to lift and accelerate a large mass from Earth's surface into space requires a large and powerful engine, and large engines require large quantities of fuel.

With respect to the Apollo lunar program, the Saturn V launch vehicles weighed in at some 2,800 tons and were able to lift a 120-ton payload into low Earth orbit, of which 45 tons were then directed towards the Moon. Designed in the 1960s under the direction of Wernher von Braun, the Saturn V rocket still holds the record for the most powerful, tallest and heaviest rocket to be successfully launched.

For all of the Saturn V rocket's lifting power, however, the R = (initial ground mass)/(payload mass) ratio was of order 62, and it still took 3 days to travel from Earth to the Moon – a distance of some 380,000 km. Although clearly an impressive engineering achievement, in many, many ways, the Saturn V liquid-fueled engine approach is clearly not the way to place large mass payloads into space, nor is it the way to initiate interplanetary, let alone interstellar, travel. In 1968, ahead of its time, an article in the magazine *Physics Today* by Freeman Dyson noted that even the most powerful chemical propulsion engines can only generate exhaust velocities of order 3 km/s, and that n rocket stages are required to reach a speed of $3n$ km/s. Given that each rocket stage adds a factor of 4 to the total ground mass, Dyson argued that the ground mass to final mass ratio R will increase as 4^n; with an $n = 5$ stage landing on the Moon and return rocket system the R value will be of order 1,024. With this result, Dyson noted that "These numbers show that chemical propulsion is not bad for pottering around the earth, but it is very uneconomical for anything beyond that."

Dyson was, of course, at the time of the Saturn V development and Apollo lunar landings closely involved with the (still mostly) classified Project Orion program sponsored by General Atomic and the U. S. Air Force. Indeed, Project Orion grew directly out of the wartime atomic bomb program, and the basic idea was

[37] The escape velocity is the speed required to just escape, without ever falling back, the gravitational attraction of a massive object. Physically it is the speed for which the sum of the kinetic energy and the gravitational potential energy of an object is exactly zero.

for a spacecraft to literally fly upon the very edge of the shock front generated by the sequential detonation of multiple nuclear bombs. The point of such a system, of course, is to use the great energy entrained within the blast wave of an atomic explosion to lift a massive payload into space. Although no such spacecraft was ever constructed,[38] the basic working plan was to lift payloads weighing as much as several million tons into low Earth orbit. Furthermore, the single-stage to orbit system could also deliver a sustained acceleration, with the eventual cruising speeds of order 1,000 km/s being attainable, making it an ideal spacecraft for an interstellar mission.

In a speculative talk given in September of 1959, Freeman Dyson discussed the possibility of transporting, via an Orion spacecraft, a colony of several thousand people to α Centauri.[39] He estimated that something of order 50 million hydrogen bombs would be required to accelerate and then decelerate the craft on a 150-year journey. Ingeniously, Dyson also suggested that the spacecraft's pusher plate might be made of uranium, so that during the progress of the voyage the absorption of neutrons (generated by the nuclear detonations) would gradually produce plutonium. Upon arriving at α Centauri, and finding a supposed habitable planet, the plutonium enriched pusher plate could then be dismantled and used to build nuclear reactors that could provide power for the new colony.

Although Project Orion was canceled in the mid-1960s, as a direct result of nuclear test ban treaties and non-proliferation agreements, other similar such programs have been considered. Project Longshot was developed in the late-1980s at the U. S. Naval Academy as an interstellar mission to α Centauri, the aim being to send a science package to the nearest star on a timescale of 100 years. The spacecraft would be powered by a pulsed fusion micro-explosion drive – an engine that was initially developed in the mid to late-1970s as part of the British Interplanetary Society's

[38] A design for a nuclear propelled vehicle, however, was patented (in England) by the U. S. Atomic Energy Commission in 1960 (British patent #877,392).

[39] See George Dyson's book, *Project Orion – The True Story of the Atomic Spaceship* (Henry Holt and Company, New York, 2002). See also the article by Freeman J. Dyson, "Interstellar Transport" (*Physics Today*, October, 1968).

Project Daedalus.[40] Superficially similar to the Project Orion atomic bomb drive, the fusion drive developed for the Daedalus spacecraft envisions the firing of high energy particle beams at small fusion pellets. The fusion pellets would contain a deuterium (D) and helium-3 (^{3}He) mixture, and the power to drive the spacecraft would be generated through the fusion reaction: $D + {}^3He \Rightarrow {}^4He + H + \text{energy}$. The energy generated through the short pulse of fusion reactions would convert the pellet casing into a highly conductive plasma ball, which would then be directed, via a constraining magnetic field, out of the spacecraft as a high velocity exhaust.

The micro-explosion fusion drive has many advantages over the Project Orion nuclear bomb drive, being physically lighter, potentially more efficient and free of any radioactive pollution products. The Daedalus drive offers great promise as a means of powering future interstellar spacecraft, although it is still currently a concept that is at least many decades away from even prototype development. In their forward to the final Project Daedalus report (published in 1978 – Ref.[41]) Alan Bond and Anthony Martin commented that the program was intended as a proof of feasibility exercise, "to establish whether any form of interstellar space flight could be discussed, in sensible terms, within established science and technology."

We are now at a crossroads. The proof-of-concept studies all seemingly indicate that on a timescale of perhaps a century from the present un-manned interstellar space missions should be entirely possible. What is needed now is the political will to fund the basic development costs. As to a time when human interstellar space travel might begin the future is entirely opaque, and we are likely many centuries, if not millennia, away from initiating such ventures.

---

[40] See, K. F. Long and P. R. Gales (Editors), *Project Daedalus: Demonstrating the Engineering Feasibility of Interstellar Travel* (a comprehensive collection of re-published papers by the British Interplanetary Society – available from www.bis-space.com/).

[41] See note 40. The British Interplanetary Society launched Project Icarus in September 2009 as a follow-on initiative to Project Daedalus, "to motivate a new generation of scientists in designing space missions that can explore beyond the solar system". Further details can be found on the website www.icarusinterstellar.org.

Between 1969 and 1972 12 humans, "The Dusty Dozen," spent a total of 22 h walking on and exploring the Moon's surface. Humanity has gone no deeper than this distance, some 380,000 km, into interplanetary space since. In his 1968 *Physics Today* article relating to interstellar transport, Freeman Dyson suggested that the first interstellar mission involving humans might be launched 200 years hence, circa the mid-twenty-second century.

This timeframe seems perhaps overly optimistic, but it is much more constructive in tone than the comments made by American physicist Edward Purcell (Nobel Prize winner for Physics in 1952) who, in 1962, argued that the very idea of human interstellar space travel "belongs where it came from, on the cereal box." The formidable Estonian astrophysicist Ernst Öpik echoed Purcell's perspective in a 1964 publication,[42] where he argued that the interstellar ramjet mechanism[43] as described by Robert Bussard in the 1960s is, "impossible everywhere..... it is for space fiction, for paper projects – and for ghosts."

Likewise, Gerardus 't Hooft (Nobel Prize winning physicist for 1999), in his less than inspiring but level-headed book *Playing with Planets* (World Scientific, 2008), further argues that humanity will not, even in a million years, travel beyond the boundary of the Solar System. Indeed, 't Hooft sees no likelihood of any biological entity, human or otherwise evolved, ever traveling into interstellar space – implying at least a simple solution to the Fermi Paradox that no aliens have ever visited Earth since they never left their home planet.

With the withering criticisms of Purcell, Öpik and 't Hooft, sage and distinguished scientists all, echoing in our ears, we are reminded of Arthur C. Clarke's famous laws concerning prediction. Specifically, his first law states that, "When a distinguished scientist indicates that something is possible, then they are probably right. When, in contrast, they suggest something is impossible, then they are probably wrong." In accordance with Clarke's first law, therefore, we presently maintain some high degree of hope that the first dedicated interstellar spacecraft will be leaving our Solar System astern some time within the next century.

---

[42] See, Ernst Opik, "Is interstellar travel possible?" (*Irish Astronomical Journal*, **6**, 299, 1964).

[43] A highly readable general text concerning the design and function of interstellar spacecraft is K. F. Long's book, *Deep Space Propulsion – A Roadmap to Interstellar Flight* (Springer New York, 2012).

3.8 Attention Span

"Civilization is revving itself into a pathological short attention span" – so wrote Stewart Brand, one of the co-founders of the Long Now Foundation,[44] in an essay first published, as the new millennium approached, in 1998. Indeed, for all of our supposed connectedness via Twitter, Facebook and the Internet we seem to be saying more and more about less and less, our wondering focus flitting from one ephemeral topic to the next.[45] It seems clear that besides the development of appropriate technologies, one of the biggest challenges that future interstellar mission planners will have to face is how to keep their missions in human memory.

Humanity, in general, does not have a particularly good record in either supporting or maintaining long-term research projects and/or businesses ventures. With innovation and change being the great engines of our society, there is little hope of finding at the present epoch continuing public, political or industrial support for space missions that will last multiple decades or centuries. The science community will also face challenges in justifying such projects – projects that will not return tangible results (publications and new science) on timescales relating to departmental reviews, promotion granting committees and funding agencies. Indeed, experimental scientists by their very nature tend to move as rapidly as possible from one experiment and field to another, deliberately avoiding very long term experiments.

The reasons for this are clear enough (and articulated above). The longest continually running physics experiment (besides that of the universe itself) appears to be that of the pitch drop study initiated at the University of Queensland, Australia, in 1927. The experiment concerns the viscosity of bitumen and began by allowing a sample of bitumen to settle in a glass funnel for 3 years. After this time, in 1930, the funnel tube was opened and then the entire cone placed over an open beaker. The first drop of bitumen fell from the funnel into the beaker in December 1938. The eighth

---

[44] See the extensive and detailed Long Now website at Longnow.org/.

[45] Nicholas Carr, *The Shallows: What the Internet Is Doing to Our Brains* (W. W. Norton and Company, 2011).

drop of bitumen fell on November 28, 2000. Truly this is not an experiment that has any interest to the public (or most scientists), but it is an experiment, nonetheless, that has run successfully for over 80 years. Perhaps, however, the pitch drop experiment offers an extremely useful guide to planning an interstellar mission – keep it extremely simple, keep it out of sight of the public and make sure that it requires virtually no human monitoring or maintenance.

Change is apparently inevitable, and change both inspires and stifles innovation. Few businesses last more than a handful of years before they are brought out or go under entirely. A study published by the Bank of Korea in 2008 found that in a survey of 41 countries from around the world, only 5,586 businesses could trace their beginnings back more than 200 years. Additionally, of those companies that were more than 100 years old, the great majority employed fewer than 300 people at any one time.

In terms of records, it appears that the oldest running family businesses is that of the Kongo Gumi company in Japan, which for over 40 generations, from 578 to 2006, specialized in the construction of Buddhist temples. When it was finally taken over and removed from family ownership the company employed fewer than 100 people. The oldest surviving corporation in Europe is that of the Swedish mining company Stora Kopparberg, which has been trading since at least 1288. After this, one of the next longest running, family owned businesses is that of the Marchesi Antinori winery – founded in 1385; the estate has been successively operated for 26 generations, and it currently employs some 400 workers.

The message seems reasonably clear, that while big industries and large employers are in the business of driving change and maximizing investor profits, they are not in the business of supporting long term non-profit-making projects. Not only this, large industry is founded on the very concept of built-in obsolescence: the products they sell, by deliberate design, mustn't be so well constructed that they operate or work efficiently for too long. This, of course, forces people to buy new products once their current ones become obsolete.

Likewise, government institutions fare little better than big industry in the long-term investment stakes.[46] Programs are cut when budgets are tight, and one can easily imagine that an interstellar program, well into its quiet, multi-decade cruise phase, would be an easy target for closure. Such a decision would be even easier for non-involved bureaucrats to make if the founders of a specific program had either retired or passed on.

Again, the message seems reasonably clear. For an interstellar mission to succeed it should not take much longer than a century to complete its mission. Preferably, it would seem, upon the basis of past human history and behavior, an interstellar mission should be designed so as to return its science results as soon as is possible, and the program should be overseen by as small a team of managers and expert personnel as possible. Likewise, human history and behavior would appear to indicate that the funding for the entire mission must be locked in place from the very outset, and that the mission itself should not be predicated upon the basis of continuous public interest and/or attention.

Although the latter should presumably be hoped for, it is difficult, given our present societal makeup, to see how it could be guaranteed. It would seem that not only must many technological barriers be overcome before interstellar exploration becomes possible, but so too must many of the current procedural practices of government, business and society be re-assessed and completely revised. A continually growing economy will not help in the advancement or initiation of interstellar exploration, such growth being entirely ephemeral and entirely inward looking (the mere chimera of accounting dreams). Likewise the ballooning growth of the human global population must be solved. As Thomas Malthus wisely wrote in his classic "An Essay on the Principle of Population" (published in 1798), there will come a time, in the not far distant future, that if nothing is done almost immediately, there will simply be too many humans for the world to support and feed. Just growing the economy (and hiding behind make-believe economic indicators such as the GDP and the like) has

---

[46] *Pioneer* 6 was launched into a circular solar orbit in December 1965 and holds the current record for longest functioning satellite – 35 years. Designed to measure solar wind characteristics and cosmic ray fluxes, it is presently not known if the satellite is still functioning; it was working when last contacted in the year 2000.

historically never helped the majority of humanity, and to suppose that some solution to feeding the global masses will always arise is no more than foolish hubris. The promise of interstellar travel is not a solution to humanity's growing list of near-term problems (increasing hunger, land loss, global warming, overpopulation, freshwater shortage, fish stock failures, and chronic pollution). Is humanity up to the task of saving itself first and then initiating interstellar travel? Only time will tell. What is certain, however, is that the business as usual plan is not an option that will carry us to the stars.

Whether or not we can envision the changes that will be required of society, industry and government to move us all successfully into the future, change may nonetheless occur. Whether one is optimistic or pessimistic about the future prospects for current humanity (and each camp appears to be well stocked with its vocal supporters), one cannot but temper the future in terms of Olaf Stapleton's masterful work *Starmaker*. Written in 1937, on the eve of the outbreak of the Second World War, Stapleton, a philosopher by training and career, takes the reader on an imagined out of body journey through a two-billion-year cosmic history in the classic manner of Kepler's (1636) *Somnium*. The disembodied narrator that leads us through the *Starmaker* describes an incredible panorama and history: galactic life is rare, intelligent civilizations even rarer, but throughout the story there is a theme of developing unity – a progressive development of unity within and between different civilizations. Here, perhaps, is our solace and most deep-rooted reason to reach for the stars.

3.9 A Series of Grand Tours

The *Voyager 1* and *2* space probes, both launched in 1977, took advantage of specific planetary alignments to proceed, via multiple gravitational slingshot encounters, upon two grand tours of the Solar System's Jovian planets. Both are still moving onwards, out into interstellar space, and *Voyager 1* holds the record for being the most distant manmade object from Earth. Jules Verne envisioned a somewhat similar planetary grand tour in his 1877 work *Off on a Comet*. This otherwise less than inspiring piece of science fiction literature saw its various *persona dramatis* take a journey from Earth, to Mercury, to Mars and then around Jupiter, eventually returning to Earth and safety. Verne's story is more about a struggle for survival than astronomy, but he did attempt to bring out some sense of scientific discovery through the dialog and the observations of the ever-irascible Professor Palmyrin Rosette. Verne's story, however, fails to inspire on many levels.

In contrast, Edmund Cooper's 1959 work *Seed of Light* is an altogether better portrayal of a grand tour and journey to the stars, the inhabitants of the giant starship Solarion searching, in vain as it turns out, for a habitable planet to colonize. The Solarion travels from the Solar System to α Centauri, then on to Sirius, then to Procyon, then to Vega, Formalhaut, Arcturus, Pollux, Achener, Regulus and Capella.

After a thousand years of exploration, the technicians aboard the Solarian develop a device, called the cosmometer, which enables travel between parallel universes. On its initial run, the cosmometer transports the Solarian and its crew back to our Solar System, selected as the best location in which to find a habitable planet! The journey ends where it began, but it also ends before it began in the sense that the cosmometer brings the spaceship back to Earth at a time set some 50,000 years into the past. Cooper's underlying message is clear: don't mess up the Earth. The story also reminds us of the fact that although modern astronomical research has revealed that planets are very common within the galaxy, a habitable planet is an altogether rarer entity.

A reverse, 'going home' grand tour through the stars is described by Isaac Asimov in his 1986 *Foundation and Earth*, and while the exact path is not readily discernible from the text, the journey essentially brings its human hero back to a forgotten, heavily polluted and entirely dead Planet Earth.

At this stage we leave the future prospects for humans living in and exploring interstellar space to the science fiction sages. We need not turn off, however, our collective imagination. Let us be bold, and let us also be brazen, and here assume that a means of propelling a small space probe at one fourth the speed of light is developed within the next half-millennium. With such speeds available to mission planners, what science motivated experiments might such spacecraft be tasked to perform?

One of the most obvious tasks that any interstellar spacecraft might carry out is that of the direct study of the interstellar medium. This would entail the measurements of its composition, its dust environment, the cosmic ray flux and the local magnetic field. Additional studies of the outer Solar System, Kuiper Belt and Oort Cloud objects could be conducted in the early stages of any mission. The final mission stage will naturally see the study of a new star system along with its potential new planets, moons, comets and asteroids. Such a journey would indeed be a wonder, a veritable *immram* of epic proportions.

Exploration of the solar neighborhood can, and presumably will, take many different forms. With the technology to construct and mount interstellar space missions in place, a space probe might be sent to a single star, or it might be targeted and programmed to visit a specific sequence of stars over many decades or several centuries. Let us assume here, for the sake of setting some limit, a mission time of less than 65 years – a good human lifetime. With this time limit and our assumed quarter of light speed space probe, then any star within the solar neighborhood, out to, say, 12.5 light years could be visited and the data returned to Earth within our time limit.

Given such constraints, the question becomes, which stars should the space probe visit and why. Although it makes obvious sense to visit α Centauri first, since it is the closest and thereby most accessible nearby star, it is worth briefly reflecting on how the nearby stars might be rated with respect to criteria other than distance. The establishment of selection criteria and object weighting factors is always a controversial exercise, since different research fields have different priorities. From a purely astronomical perspective, however, we suggest a stellar weighting scheme something like that displayed in Table 3.2 might be useful. The scheme envisioned relies on the evaluation of four parameters: $W = R_{dist} + R_{NSP} + R_{planet} + R_{disk}$. The first term is a weighting factor for distance, with $R_{dist} = (12.5 - d)$, where d is the distance to the system in light years. The 12.5 corresponds to the maximum distance that the probe can travel and return mission data to Earth within our set time limit of 65 years or less.

Table 3.2 Weighting factors arranged according to spectral type and object designation. Column 2 indicates the number of stars, with the specified spectral type, observed within 5 pc of the Sun (Data from Cantrell[a]). Column 4 indicates the percentage of stars, of the specified spectral type, that are observed to have planets (Data from Johnson[b]) (*) The Sun is included in this total

Object	N_{SP}	R_{NSP}	$N_{P\%}$
O	0	10.0	0
B	0	10.0	0
A	1	9.86	10
F	1	9.86	10
G	3*	9.58	5
K	7	9.03	5
M	50	3.06	2
WD	5	9.30	7.5
BD	5	9.30	0

[a]Justin Cantrell, T. Henry and R. White, "The Solar Neighborhood XXIX: the habitable real estate of our nearest stellar neighbors" (arxiv.org/pdf/1307.7038.pdf). See also Note 40 and specifically the paper by Anthony Martin, "Project Daedalus: the ranking of nearby stellar systems for exploration."

[b]Details provided in the review paper by John Johnson in the *Publications of the Astronomical Society of the Pacific* (**121**, 309, 2009). The numbers shown in Column 4 of Table 3.2 are clearly going to change over time, since our knowledge of planet occurrence rates is currently far from complete. The percentage number of white dwarf objects having associated planets is very uncertain at the present time, and we set it to 7.5 % on the basis that this is the characteristic number associated with the A to K spectral type stars – which are the essential progenitors of white dwarfs. We have simply assumed that brown dwarf (BD) objects do not harbor planets, but this is far from clear, and this number may need to be adapted pending further survey results. Likewise, we assume that planets are not likely to form around O and B spectral-type stars. The data for Column 2 is not going to change any time soon, and with the exception of brown dwarf objects the numbers are taken to be definitive (at the present epoch)

The R_{dist} weighting factor is set up so as to favor the targeting of the closest stars. The second weighting factor is actually related to the number of stars, N_{SP}, of a given specific spectral type, found within 5 pc (16.3 light years) of the Sun, and we set $R_{NSP} = 10\,(1 - N_{SP}/72)$. The data for the N_{SP} term is shown in Column 2 of Table 3.2, and these correspond to the total number of actual stars, of a specific spectral type, within 15 light years of the Sun. Indeed, in such a distance-limited survey there are 62 stars (including the Sun), and 5 white dwarfs (WD) in 50 systems, comprising 34 single stars, 11 binary stars and 5 triple-star systems. There are also 5 known brown dwarfs (BD) within 15 light years of the Sun, making for a total of 72 objects within our sphere of interest. The R_{NSP} weighting factor has been constructed so as to favor the less common spectral type stars, along with the white dwarf and brown dwarf objects.

The third weighting factor R_{planet} relates to the prospects of finding a planet, or planetary system at the targeted star. For this term we setup a conditional statement related to the present-day observations. If a star is known to harbor a planet or planets then $R_{planet} = 10$ x number of known planets. Alternatively, if a star has no observed planets then R_{planet} is set equal to $N_{P\%}$ as given in Column 4 of Table 3.2. This latter data is taken from the review paper by John Johnson (Institute for Astronomy, Hawaii)[47] and describes the percentage of stars, of the specified spectral type, that are observed to harbor planets. The main aim of this term is to favor the exploration of systems harboring known numbers of planets. No specific adjustment to the R_{planet} weighting factor is applied with respect to the kind of system in which a star is found. This might be important, for example, for those stars located within close binary systems where the region within which they might form planets could be restricted.

---

[47] See, John Johnson's review paper in the *Publications of the Astronomical Society of the Pacific* (**121**, 309, 2009). The weighting scheme set up in Table 3.2 places no specific value on the habitability of the planets (recall Sect. 2.16) that might be located within a specific system. This factor could, of course, be introduced if desired, and for example this was the basis of the weighting scheme developed by Anthony Martin for the British Interplanetary Society's Project Daedalus (see Note 40). Accordingly factors relating to the probability of a given type of star and planet configuration being able to support autotrophic life might be considered. The presence of terrestrial planets is typically taken as a minimum requirement for a system to be considered as potentially habitable. This, however, does provide a specific bias that might not always be justified. Indeed, in his recent book, *The Beginning and the End: the meaning of life in a cosmological perspective* (Springer, 2014), Clémet Vidal has suggested that advanced life may choose to transfer its 'essence' to a post-biological substrate which can 'feed' directly upon the energy supplied by a star. Vidal introduces the term *Starivore* to describe such life forms, and suggests that such civilizations might inhabit semi-detached binaries and actively exploit the mass flow and accretion phenomenon associated with such systems. Since, however, there are no semi-detached or cataclysmic binary systems within 100 pc of the Sun we do not consider them in our weighting scheme.

The final weighting factor R_{disk} relates to the observation, or not, of an extended dust disk about a given star. Here we simply set $R_{disk} = 2 \times$ number of observed disks if a disk(s) has been detected[48]; otherwise, $R_{disk} = 0$.

The weighting factors that we have adopted above could easily change in future years. They are an attempt, however, to illustrate the processes that might be used to select target stars for future interstellar missions. A few example determinations are called for. For the Sun, for example, we have $d = 0$, and accordingly $R_{dist} = 12.5$. Given the Sun has a G2 spectral type, so $R_{NSP} = 9.58$. The Sun has 8 observed planets, giving $R_{planet} = 80$, and finally, the Sun has an Asteroid Belt as well as a dust disk associated with the Kuiper Belt, indicating that $R_{disk} = 4$. The total weighting factor for the Sun is accordingly W = 105.08. For τ Ceti, we have a distance of $d = 11.9$ ly, giving $R_{dist} = 0.6$; τ Ceti is a spectral-type G8.5 star, and accordingly, like the Sun, it has $R_{NSP} = 9.58$. There are 5 planets in orbit around τ Ceti, which gives $R_{planet} = 50$, and an extended dust disk has been detected around the star, so $R_{disk} = 2$. The final weighting factor for τ Ceti, therefore, is W = 62.2.

For a final example we look at the system Sirius AB. The distance to Sirius AB is 8.6 ly, and so $R_{dist} = 3.9$ for each star. The two components making up the Sirius binary are an A1 star (Sirius A) and a white dwarf (Sirius B), and accordingly, $R_{NSP} = 9.86 + 9.30 = 19.16$. No planets are known about either star, so, from Column 4 of Table 3.2, $R_{planet} = 10 + 7.5 = 17.5$. No disk-like structures have been observed in association with Sirius AB, so $R_{disk} = 0$. Summing all the terms together gives a final weighting factor of W = 44.46 for Sirius AB. Table 3.3 shows the top 10 stellar systems ranked according to their weighting factors.

---

[48] The presence of a disk is typically betrayed through infrared and microwave wavelength observations, where it is the thermal radiation from the dust that is detected. Sometimes the extended disks are observed optically, as in the classic case of β Pictoris (Fig. 2.11) and with the bright star Formalhaut (Fig. 3.4). The size of the disk provides some information about its origins, in the sense that if, for a Sun-like star, say, the disk has a radius greater than 50 au, then the dust is most likely derived from the collisions between icy Kuiper Belt analog objects. An example of this type of disk is that observed around the bright star Vega. If the disk has a radius of just a few astronomical units then the dust is probably associated with a population of stony/iron asteroid-like objects. An example of this kind of disk is that observed around τ Ceti.

TABLE 3.3 Top ten stellar systems ranked according to their associated weighting factor W. Columns 4 and 5 indicate the presently observed number of planets and disk components. A question mark appears in the planets' column of ε Eridani, since it is highly likely that the observed dust/debris disk structures could only be maintained through the actions of shepherding planets

System	Components	*d* (ly)	Planets	Disks	W
Sun	G2	0.0	8	2	105.08
α Cen	G2, K0, M5.5	4.23	1	0	63.04
τ Ceti	G8	11.86	5	1	62.18
Sirius AB	A1, WD	8.60	0	0	44.46
Procyon AB	F5, WD	11.41	0	0	38.85
ε Indi	K4, BD, BD	11.83	0	0	34.91
61 Cyg AB	K5, K7	11.41	0	0	30.19
Luhman 16 AB	BD, BD	6.58	0	0	24.80
ε Eridani	K2	10.50	?	3	22.01
L 789-6	M7, M5, M5.5	11.08	0	0	19.13

The top ranked star in Table 3.3 is the Sun, and this is exactly as it should be, given that it is the closest star to us and that it hosts eight known planets, an asteroid zone and an extended Kuiper Belt. The number one ranking of the Sun further acts to remind us how singularly important the Solar System is to us, and it underscores the imperative to know and explore the great riches that it offers.

The second ranked system is α Centauri, and this comes about because of the system diversity: three close, different spectral-type stars, with at least one known planet. The star τ Ceti, although situated near the upper limit of our adopted mission distance, nonetheless ranks number 3 in the list. This is primarily because of its extensive planetary system. Sirius AB and Procyon AB rank 4th and 5th in the list in relation to their being relatively close binary systems, with each component ranking highly with respect to the probability of supporting planets.

The binary systems ε Indi and 61 Cyg AB come in 6th and 7th primarily because they have K spectral-type components that rank relatively highly with respect to the probability of supporting planetary systems. ε Indi also ranks highly because of being a triple system in which two of the components are brown dwarfs. The binary brown dwarf system Luhman 16AB places 8th in the table primarily because it is a nearby system and because its two components rate highly with respect to their relative rarity (rarity, that is, on the scale of the adopted mission distance limit of 12.5 ly). The star ε Eridani comes in 9th because of its K spectral type and the fact that three distinct disk structures have been observed to encircle the star. It is highly likely that there are planets within this system, and accordingly when they are eventually identified its overall ranking will greatly improve. The final, 10th ranked system L789-6, while again close to the 12.5-light-year mission limit adopted rates relatively highly because it is a triple-star system.

Sitting just outside of the selection range for Table 3.3, at a distance of 12.78 ly, is Kapteyn's star. This interesting red dwarf star was initially recognized for its high proper motion. Indeed, it is second only to Barnard's star (recall Fig. 1.18) in terms of its rate of motion across the sky. First cataloged by Dutch astronomer Jacobus Kapteyn in the late nineteenth century it was subsequently recognized as being a so-called galactic halo star. In this respect its motion is retrograde, that is, opposite to the spin of the Milky Way's disk, and it has a large eccentric orbit and a high radial velocity.

Intriguingly, Kapteyn's star was most likely born in an entirely different galaxy to our own, and was only later accreted (along with the rest of its parent dwarf galaxy) by the Milky Way. The estimated age of Kapteyn's star, at about ten billion years, makes it more than twice as old as our Sun and indicates it must be one of the very first formed stars in the universe. Interest in this system spiked in June 2014, when Guillem Anglada-Escudé (Queen Mary University of London) and co-workers announced the discovery of two associated planets: one planet has an orbital period of 48.616 days and a minimum mass of 4.8 Earth masses, while the other has a period of 121.54 days and a minimum mass of 7 Earth masses. As shown earlier in Fig. 2.26, the innermost planet (Kapteyn b) sits close to the outer edge of the habitability zone, and this suggests it could be a highly important astrobiology target. With its associated planets, Kapteyn's star, in spite of our distance weighting scheme, still attains a healthy W = 22.7 rating, technically placing it above ε Eridani in Table 3.3. Although Kapteyn's star made its closest approach to the Solar System some

11,000 years ago, it is still the closest star to us with a known planet (a super-Earth) located within, or close to, the system's habitability zone.

Since different weighting systems to the one described above can be envisioned the ranking shown in Table 3.3 is not the only one, but it is unlikely that the ranking of the first three systems will change in the future.

Clearly, at the present epoch, the Solar System is the most important system for us to be exploring. With the advent of interstellar spaceflight capability, however, it seems reasonably clear that α Centauri should be the first target system. Indeed, it is an absolutely ideal target for the first interstellar mission. Not only is it the closest star system to us, but it is composed of three stars having distinctly different spectral types, and the system contains at least one planet. We might further expect the weighting factor for α Centauri to increase in the next several decades, firming up its second place ranking, as a result of more planets being discovered in the system.

After α Centauri, then the next most interesting system with respect to the number of known planets is τ Ceti. The travel time to and return of science data from this star, however, is relatively long; indeed, it is close to the 65-year limit adopted for our mission profile. Given this limitation, the astronomical ranking system, as used to formulate Table 3.3, could be abandoned and a simple distance-related weighting scheme adopted. Accordingly, after α Centauri, the next closest target systems would be Barnard's star (otherwise ranked 14th) and Luhman 16 AB (otherwise ranked 8th).

The low ranking of Barnard's star in our scheme is perhaps worth commenting on, since it was the number one ranked star in the British Interplanetary Society's Project Daedalus mission profile. Their high ranking came about because it was once thought, erroneously as we now know, that there might be two planets in orbit around the star (recall Sect. 1.17). If Barnard's star did indeed have two planetary companions, then it would rank 8th in our list (shown in Table 3.3). This ranking is still on the low side, but comes about because of our relatively low weighting ascribed to M spectral-type stars.

So far we have considered one-off missions. That is, the space probe is targeted to study one specific star system, and to return science data back to Earth in a mission time shorter than 65 years. Such a time limit, of course, is arbitrary and there is no specific reason why 100 year and longer missions might not be considered. Additionally, the constraint of visiting just one stellar system can be relaxed, and in this latter case one can envision a space probe being sent on a grand tour of stellar systems within the solar neighborhood.

At this stage it is worth taking a brief reminder-look at Fig. 1.18, which shows the three-dimensional distribution of stars out to a limit of about 15 light years from the Sun. Even without any specific numbers it is reasonably clear from Fig. 1.18 that there is no simple path that a space probe might follow in order to systematically visit all the stars in our inner solar neighborhood. Once again, therefore, some form of ranking or simple rule-based system might be adopted to guide in the targeting of specific systems for a grand tour route. Fig. 3.8 shows four grand tour journeys that future interstellar space probes might be sent on with respect to stars located within 5 pc of the Sun. The routes have been determined according to a specific set of rules, and these are: Tour 1 corresponds to the shortest possible path between neighboring systems visited; Tour 2 corresponds to the greatest diversity of objects visited enroute; Tour 3 corresponds to a tour that incorporates a visit to τ Ceti; and Tour 4 corresponds to a tour that incorporates a visit to ε Indi. The system diversity and total travel distance associated with each tour is shown in Table 3.4.

Tours 1 and 2 start along the same path, first visiting α Centauri and then heading for the binary brown dwarf system Luhman 16 AB, located about 1.1 pc (3.59 ly) away. After this encounter, Tour 1 sees the interstellar probe head towards the M9 dwarf star Denis 1048–39, followed by flybys of the M5.5 dwarf star L 143–23 and the white dwarf LL 145–141. Tour 2, in contrast, heads to Sirius AB after its flyby of Luhman 16AB,[49] with subsequent visits to Procyon, Luyten's star, DX Cancri, Lalande 21185, Wolf 359,

[49] After visiting Luhman 16AB, our imagined Tour 2 interstellar spacecraft could visit brown dwarf object WISE J085510.83-071442.5 (hereafter WJ08). Technically this object is closer than Sirius, but at present there is little reason to suppose that WJ08 is an object significantly different from the individual members that constitute Luhman 16 AB, and there is no reason to suppose that WJ08 might harbor any planets. The end of Tour 2 could be reorganized, however, to visit WJ08 after the flyby of Wolf 359 and prior to heading onwards to Ross 128.

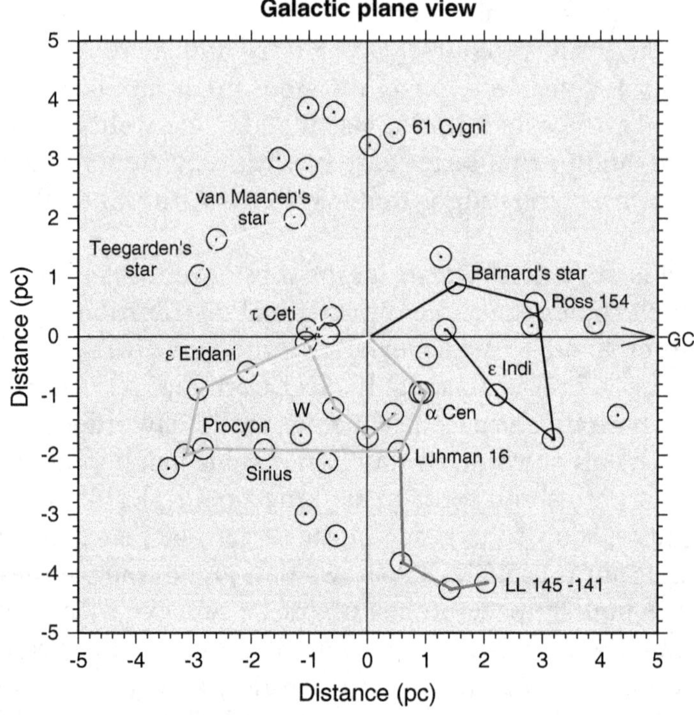

FIG. 3.8 Four grand tours that future interstellar space probes might be sent upon. The various voyages while moving through three-dimensonal space are here projected onto the galactic plane. Tour 1, the shortest distance between stars voyage, is shown by the *red line*. Tour 2, the greatest diversity route, is shown by the *green line*. Tour 3, the grand tour, including a stop-off at τ Ceti, is shown by the *yellow line*. Tour 4, the voyage that includes a visit to ε Indi, is shown by the *black line. W* indicates the location of the recently (2014) discovered brown dwarf WISE JO85510.83-071442.5

TABLE 3.4 System diversity and total travel distances for four possible grand tours within the inner solar neighborhood. WD = white dwarf, BD = brown dwarf

Tour	Systems visited	Distance (ly)
1	5 stars, 1 WD, 2 BDs, 1 planet	22.33
2	12 stars, 2 WDs, 2 BDs, 1 planet	46.66
3	7 stars. 1 WD, 1 BD (?), 5 (+?) planets, 1 debris disk	34.69
4	5 stars, 3 BDs,	29.88

Ross 128 and Wolf 424AB. After visiting Procyon none of the hops in Tour 2 are longer than 1.7 pc (5.54 ly) in length, and indeed, the journey between Luyten's star and DX Cancri is just 0.35 pc (1.14 ly). The one known planet that will be encountered during Tours 1 and 2 is that of α Cen Bb.

Tour 3 begins by traveling to ε Eridani, a journey of some 3.22 pc (10.50 ly) distance where, upon arrival, the space probe will encounter the star's poorly understood (at the present time) debris disk and possible planetary system. The existence of planets is quite likely since the inner two of the debris belts appear to be quite distinct, and the most likely way of achieving this dichotomy is through the gravitational shepherding of planets. The (+?) term in Table 3.4 has been used to indicate this possibility. From ε Eridani the space probe travels on to the binary system UV Ceti and then to τ Ceti and its associated five-component planetary system. After visiting the M5 dwarf star YZ Ceti, Tour 3 continues by conducting flybys of Van Maanen's stars – a white dwarf with a possible brown dwarf binary companion, L 1159–16 and Teegarden's star, the latter two objects both being M dwarf stars.

Tour 4 is mostly about visiting M dwarf stars. The first leg of the journey sees the space probe travel to Barnard's star. Ross 154 is the next star encountered, followed by SCR 1845–6357, which is a binary system composed of an M8.5 star and a brown dwarf. The tour then moves towards the triple system ε Indi, composed of a K4 star and two brown dwarfs, and thereafter onwards to an encounter with the M2 dwarf star Lacaille 9352.

From an astronomical perspective no specific distinction is drawn between the four grand tours listed in Table 3.4. They all enable the study of a diverse set of stars and degenerate objects, and they are all constructed on the basis of selecting the smallest distances between neighboring systems. In terms of overall mission time, and for a given space probe velocity, then Tour 1 returns its data in the shortest amount of time, followed by Tours 4, 3 and 2, respectively. Tour 2, however, returns the greatest amount of data relating to stars, white dwarfs and brown dwarf objects. Tour 3 is biased towards returning exoplanet data (on the basis of currently known numbers), while Tour 4 maximizes the study of M dwarf stars and brown dwarf objects. The ideal option, of course, would be to construct four or more interstellar space probes – one for each grand tour, and extras for any additional routes that might be configured or deemed specifically interesting.

The tours described above take the space probes no more than 15 light years from the Sun, over time intervals falling between 89 and 187 years (assuming, recall, a quarter light speed velocity), but if longer mission times are deemed supportable then clearly larger and more distant grand tours, lasting many centuries, could be constructed.

3.10 Finding ET: Finding Ourselves?

Interstellar travel will not solve any of the humanity's current problems. It will hopefully, however, help humanity define a better future for itself, and maybe, just maybe, it will enable direct contact with life beyond the Solar System. Life is certainly a very precious commodity, and although it is clear that the essential molecular materials, the building blocks that make life possible, are available in great abundance in the interstellar medium, it is far from clear that the spark of life has been breathed into those same molecules in any world other than on our own Earth.

This is not hubris, nor faith, but just a statement of current observational facts. The four most abundant elements in the interstellar medium are hydrogen, helium, oxygen and carbon, and 99 % of the 10^{28} atoms that constitute a 70-kg human body are composed of these atoms. The remaining 1 % of atoms in the human body is divided between an additional 37 elements from within the Periodic Table (including iron, copper, tin, lead, gold, strontium and uranium). Indeed, there are not many of the 79 stable periodic elements that life, through the ever searching fingers of evolution, hasn't found a use for.[50] Not only is it remarkable, indeed wonderfully so, that an assortment of 10^{28} atoms can be

[50] The noble gas argon (Ar), the third most abundant element in Earth's atmosphere, is the one clear exception to this rule. For all this, however, Harlow Shapley has described a time and space linking connection between atmospheric argon and the accumulated breaths of all human beings that have ever lived in his engaging 1967 book *Beyond the Observatory*.

arranged and assembled into a living, breathing entity that can walk, talk and think, the whole synergistic relationship between atoms, evolution, life, the stars, and the universe is breathtaking in both its inherent beauty and complexity.

The search for extraterrestrial life (SETI) has now been going on for over 50 years, and to date the various radio telescope surveys and searches have heard nothing but a roaring silence. Nor have the optical surveys revealed the presence of any interstellar laser communications; nor have infrared wavelength surveys found any evidence for the existence of Dyson spheres or Kardashev Type II civilizations.[51] Nor have any gamma ray signals been detected to betray the use of fusion-powered interstellar spaceships. In short if ET is calling or flying home, then it is being done in a manner that so far eludes us. None of these non-detection results, of course, mean that extraterrestrial life does not exist.

A lower limit on the potential number and separation between locations where life might have evolved is provided for by looking at the potential number of habitable planets within the Milky Way Galaxy. It is a matter of choosing one's anticipated numbers and calculating the odds. And, although this calculation is partly a fool's errand, since many of the numbers are guesses at best, it does enable us to focus on the unknowns as well as the (partially) known requirements for a galactic civilization to even potentially exist. And we can in principle catch our first glimpses of where and how humanity and our resplendent Earth fit into the bigger picture.

The idea of the habitability zone has been discussed in several sections already, but when journeying into the greater galaxy, it is the habitability of a much larger zone that is of specific interest here. Counter to the Copernican Principle, the Solar System, along with its very special third and inhabited planet, is not located at some random galactic point. Indeed, life cannot exist just

[51] V. Badescu and R. B. Cathcart, "Stellar engines for Kardashev's Type II civilizations" (*Journal of the British Interplanetary Society*, **53**, 297, 2000). Russian astronomer Nickolai Kardashev introduced the idea of a technology type in 1964. Three civilization numbers were introduced with labels I, II and III being assigned according to the control of materials and energy resources at the level of planetary, host star and host galaxy, respectively.

anywhere, and in terms of life as we currently recognize it (i.e., requiring a habitable planet or moon) the conditions for it to evolve, thrive and expand are limited in both time and space. In a series of research papers published from 2004 onwards, Charles Lineweaver (University of New South Wales, Sydney) and co-workers have outlined[52] the idea of the galactic habitability zone (GHZ). In this manner the habitability of the galaxy has been assessed in terms of four criteria: the presence of suitable host stars, the presence of enough chemical elements to enable the formation of planets, the time required for life to evolve, and the presence of non-life-exterminating environments.

The second and fourth conditions outlined by Lineweaver et al. present us with a threshold condition. Indeed, it is a life-giving and death-taking struggle that is invoked, and the key role is that played by supernovae. As we have seen in other sections of this book, supernovae are the great movers and shakers of the interstellar medium. They compress and they disperse, they initiate star formation and they destroy stellar nurseries. Not only this, however; they are also the great machines that nature uses to change the abundant hydrogen and helium, produced in the primordial Big Bang, into other elements of the Periodic Table of Elements. All the carbon, oxygen, sulfur, potassium and other elements essential to allowing life to come about are produced within supernovae, the destruction of massive stars. As American astronomer Carl Sagan was often heard to say, "We are star stuff."

It is by transmuting hydrogen and helium into other elements that the wonderful alchemy of the stars, and especially of supernovae, has enabled the eventual possibility of complex structures, such as life, and of course, planets, to come about. Supernovae, however, wield the proverbial double-edged sword, and though planets and life would not exist without their occurrence, life can also be wiped away by their too close detonation. We have, once again, a classic Goldilocks effect coming into play. For life to come about there must be an epoch of star formation and supernovae recycling of the initially hydrogen- and helium-rich interstellar

---

[52] The details are provided in Charles Lineweaver et al., "The Galactic Habitability Zone and the Age Distribution of Complex Life in the Milky Way" (the article is available at arxiv.org/abs/astro-ph/0401024).

medium. In contrast, for life to thrive and evolve, its relative environment must be largely supernovae free. The trade-off of these conditions determines when life might first appear in the galaxy, and it also determines where it might appear. The when part of the GHZ equation is determined by the galaxy's star formation history, which observations reveal was higher in the past than in the present. The early, intense bout of star formation is important for generating the raw materials out of which planets and life will eventually form and evolve, but the supernovae rate is far too high early on for stable, life nurturing environments to exist. With the passage of time, however, the star formation rate drops, the supernovae rate falls, and within an ever increasing number of small, quiescent galactic pockets the great engines of evolution and natural selection can grind their way to the production of life.

Lineweaver and co-workers write, "Poised between the crowded inner bulge and the barren outer Galaxy, a habitable zone emerged about 8 Gyr ago." The inner and outer boundaries for the GHZ annulus are currently set between 7 and 9 kpc from the galactic center, and interestingly, Lineweaver et al. also find that the majority (75 %, in fact) of stars in this region are about a billion years older than the Sun and Solar System. We are literally surrounded by older galactic worlds, and possibly by much more intelligent intergalactic species.

The extent of the GHZ, as presented above, can be used to estimate the likely distance between advanced galactic civilizations. If we allow the annulus of the GHZ to be 1,000 pc thick (that is, stretching 500 pc above and below the galactic plane), then the volume of space encompassed is $V_{GHC} = \pi \times 1 \times (9^2 - 7^2) \approx 10^{11}$ $(\text{pc})^3$. The total number of stars in this volume of space will be of order $N_{GHC} \approx V_{GHC} \times 0.01 = 10^9$ [where the 0.01 term is the typical number of stars per cubic parsec of space within the disk of the galaxy]. If we restrict ourselves to just Sun-like stars, then of order $N_{SL} \approx 80$ million such stars exist within the GHZ.

Current exoplanet survey results indicate that perhaps 50 % of all Sun-like stars have an associated planetary system – which gives us of order $N_{LS} \approx 40$ million potential life-supporting structures. Once again, current estimates suggest that perhaps one in ten planetary systems might contain a planet within the habitability zone of its parent star. Our adopted numbers indicate, therefore, that there should be something like $N_{HL} \approx 4$ million potential life-supporting host planets in the GHZ. If these systems are distributed at random within the GHZ, then the likely volume of space associated with each life-supporting planet is of order $V_{GHC}/N_{HL} \approx 2.5 \times 10^4$ cubic parsecs, and this indicates a typical separation distance of order 37 pc between such worlds. Taking our derived numbers as being representative, then, the odds of a life-supporting planet existing within the habitability zone of α Centauri are of order $(1/37)^3$ or of order 1 in 50,000.

Bearing in mind the famous, and very astute "lies, dammed lies, and statistics" adage of Benjamin Disraeli, our estimate for the habitable real estate within the GHZ could easily be increased if, in addition to Sun-like stars, the K and M spectral-type stars are included in the count. In this case, rather than about 8 % of the N_{GHC} stars being available to host habitable planets, nearly 90 % (or some 900 million) become potential hosts. This greater number of stars (all else being the same) reduces the potential habitable planet-hosting system separation to about 15 pc (~50 light years). Now, the probability of a habitable planet existing within the α Centauri system (extended to include Proxima Centauri) is improved to a more healthy $(1/15)^3 \approx 2 \times 10^{-5}$, or about 1 in 3,500. This latter number is no doubt on the optimistic side, but it highlights the point that a life-supporting planet need not be located at an excessively great distance from us.

At this stage we have simply attempted to calculate the possible number and average distance between star systems hosting a habitable planet. A planet being habitable, of course, need not mean that life actually evolves there, and even if life does evolve, this does not mean that it ultimately becomes intelligent life in the sense of developing abstract and higher cognitive reasoning. If just 1 in 50,000 of the available habitable planets evolves life that is capable of developing a complex, potentially spacefaring society, then perhaps 1,000 such civilizations exist within the GHZ, with the typical separation between such civilizations being of order 600 pc. The present consensus of the numbers would seem to suggest that simple life (although there is nothing simple about it) is probably common within the galaxy; intelligent life-forms, on the other hand, capable of exploring interstellar space and/or developing large astro-engineering structures would, at best, appear to be a very rare galactic occurrence.

The physical distance between potential extragalactic civilizations, even if our optimistic numbers are close to being the correct ones, presents an enormous challenge if direct contact is to be hoped for. The possibility of contact certainly improves if one allows for the colonization of the galaxy to take place; we have only considered static civilization numbers. Indeed, many reasonable colonization strategies have been outlined over the past several decades, and given just a few hundreds of millions of years time a vigorously expansionist civilization could take over a vast swath of the galaxy. There is no evidence that such galactic colonization has ever taken place, which does not tell us that intelligent space-exploring civilizations do not exist, but rather it simply tells us that exploring the galaxy is extremely hard – even if you are a civilization, as seems quite likely from the age distribution of stars within the GHZ, a billion years older than our own.

One reason why our distant descendants might ultimately make the interstellar leap could spring from the evolved imprint of our ego, which currently drives our thirst for exploration and adventure and also acts to preserve human existence. The first drive is focused on the science of discovery, while the second seeks to perpetuate and even expand life, human life, within the greater galaxy. Indeed, the very thought of our complete and utter eradication, there literally being no human alive anywhere in the entire vast universe, seems to frighten our deep psyche. We rebel against this notion, even when over-arched by the aphotic shadow of an entirely indifferent universe.

We are indeed self-aware, and we know that as individuals death is inevitable, but collectively a sense of solace is found in the promise of new generations to follow. Cut off that future promise, and we are lessened as individuals and we are weakened as a community. Accidents and natural disasters will happen, asteroids and comets will impact upon the future Earth, super-volcanoes will explode and tsunamis will break upon defenseless coastal plains. Tectonic plates will split apart and relentlessly collide; nearby supernovae will detonate, and spiral arm passages by the Sun and Solar System will occur, but we stubbornly cling to the idea that somehow, just as in the past, the human race will endure. Indeed, tenacity in sight of future disaster will be a strong motivator for interstellar migration – *per ardua ad astra*. Although human tenacity will play its part in our long-term collective survival, the words of Gerald Bostock (alter ego of musician Ian Anderson) in his recent project "Homo Erraticus" (Calliandra Records, 2014) remind us of Carl Sagan's "It will not be we" and the fact that much else will need to change about ourselves as we move into the unstoppable future. Bostock writes, "The hunter-gatherer is still amongst us. Embedded in the collective psyche....We are the angry species. The ones who soil our nest and journey to occupy another."

3.11 The Life of a Sun-like Star

We now plunge deep into the future. The motivations and activities of humanity have now passed beyond even our rampant speculation, and we enter into the glacial timescale of stellar evolution. Our time steps are now measured not in centuries but many millions and even multiple billions of years. The timescale now exceeds all that has gone before us, and we enter a time when the solar neighborhood will have changed beyond all recognition. The Sun and α Centauri will no longer be close companions, their proper motion paths and their galactic orbits having carried them to entirely different regions of the Milky Way.

What has gone before us, in the entire 4.5-billion year history of Earth, has been just the lull before the storm, and the one future event that will assuredly destroy everything on this planet, with no exceptions, from algae to zooplankton, looms. This the aging of the Sun. Indeed, the aging Sun will eventually drive life from the Solar System. If we stay we perish *holus bolus*. Just as a mother bird eventually drives its siblings from the natal nest, so future representatives of humanity must ultimately fly from Earth and find new planets and new moons upon which to live, or else humanity ends – full stop. There will likely be no exodus for the majority; these journeys, should they be made, will save only a minuscule proportion of our future descendents and the world's wonderful menagerie of life. The passing of the Sun is looming – albeit at the present time very slowly.[53]

How is it that the Sun can turn so dramatically upon all life, from nurturing mid-wife to hooded executioner and somber undertaker? Of all the possible reasons for this turncoat behavior, the rather mundane answer lies in the manner in which it generates its energy. The Sun, as we saw earlier, generates its energy through the fusion of hydrogen into helium, and this process necessarily changes the composition of the core region. It becomes increasingly helium rich and hydrogen poor. It is this change in composition, along with the limited amount of fusion fuel[54] (i.e., the hydrogen) that forces a star to change its internal structure over time. The internal changes within a star are ultimately manifest at the surface by changes in its observable temperature, size and energy output (that is, its luminosity).[55]

To briefly recap, it was argued earlier that the Sun, as well α Cen AB and Proxima, are main sequence stars, and by this it is implied that they are generating internal energy through the fusion of four protons into a helium nucleus made of two protons

---

[53] Is the deadly and life-on-Earth-destroying increase in the Sun's luminosity inevitable? The answer to this is yes, if the Sun is left to its own devices. Need humanity, or the ego of humanity, inevitably seek survival by flying through interstellar space to new stars and planetary systems? To this question the answer is both yes and no. Yes in the sense that sooner or later the Sun will perish and the Solar System's habitability zone will eventually collapse. Alternatively, no is the answer to the question if one is prepared to envisage astro-engineering possibilities. The no, of course, is a relative no in the sense that although astro-engineering might considerably extend the functional lifetime of the Solar System's habitable zone, it cannot extend the lifetime indefinitely. The author has discussed many of the proposed astro-engineering options that might be applied to extend the habitability of Earth and/or counteract the red giant phase of the Sun's evolution in the book: *Rejuvenating the Sun and Avoiding Other Global Catastrophes* (Springer New York, 2008). The author has additionally discussed the possibilities of terraforming both the planets and moons within our own Solar System and those in orbit about other stars in the book: *Terraforming: The Creation of Habitable Worlds* (Springer New York, 2009).

[54] Here, in some sense, is the ultimate example of the limitations described by Thomas Malthus in the late eighteenth century. The Sun contains of order 10^{57} protons, all brought together in a structure containing 2×10^{30} kg of matter. And yet, for all this great mass and unimaginable number of particles, it has a finite lifetime.

[55] See the author's book *Rejuvenating the Sun and Avoiding Other Global Catastrophes* (Springer New York, 2008). The reader is also referred to the comprehensive book by R. C. Smith, *Observational Astrophysics* (CUP, 1995). A detailed mathematical treatment of the subject is given by C. J. Hansen and D. Kawaler, *Stellar Interiors: Physical Principles, Structure and Evolution* (Springer-Verlag, 1994).

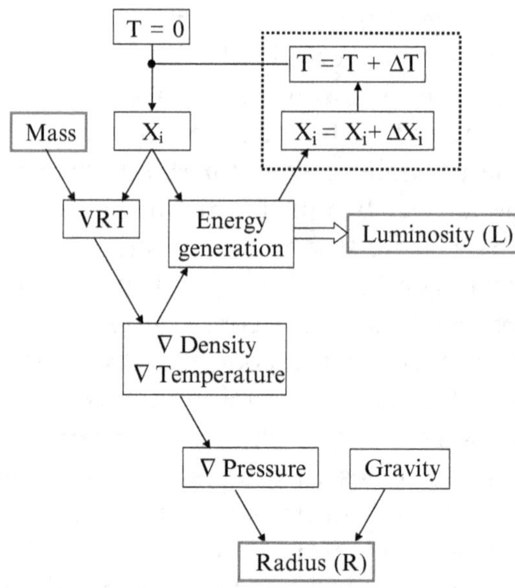

Fig. 3.9 It is the alteration in a star's internal composition, changed by the fusion reactions operating within its central core, that ultimately causes its surface temperature, radius and luminosity to change over time. The evolution driving loop (*dotted box*) accounts for the change in composition, which then drives changes in the star's internal structure. **Key**: X_i is the ith chemical element; ΔX_i is the change in the abundance of element X_i in time interval ΔT; VRT signifies the Vogt-Russell condition, ensuring a unique solution to the equations of stellar structure. The ∇ symbol signifies the generation of a spatial gradient through the interior of the star. Once the surface luminosity (L) and radius (R) are determined, the surface temperature is defined by the Stefan-Boltzmann law

and two neutrons (recall Fig. 2.3). It was also revealed earlier that the luminosity, radius and temperature of a main sequence star are uniquely determined once the mass and the internal variation of chemical composition is determined. This result is encapsulated within the Vogt-Russell theorem. Indeed, it is from this starting point that we can see why the observable characteristics of a star change over time. Schematically we have the situation shown in Fig. 3.9.

To interpret Fig. 3.9, we need to recall that in order for a star to not collapse in upon itself due to its own gravity, it must establish an internal pressure gradient. It is through this internal pressure gradient, brought about because the interior is hot, that gravity is held in check. The equilibrium is dynamic, however,

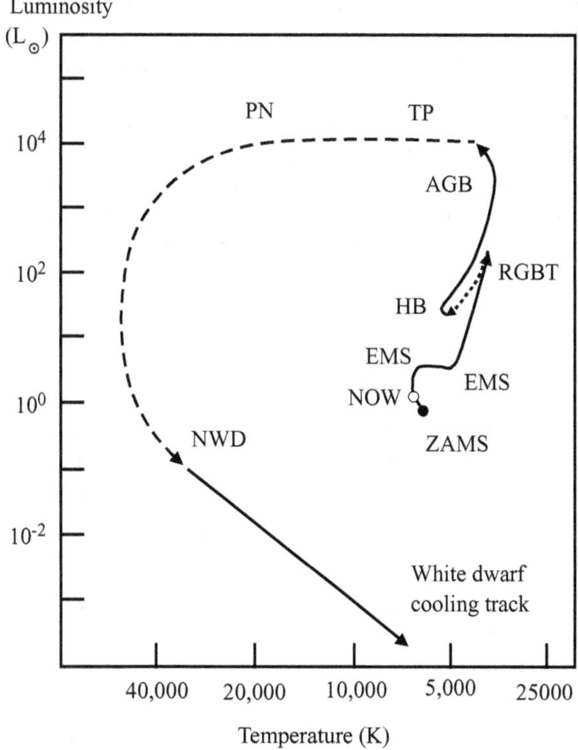

FIG. 3.10 A schematic diagram of the Sun's evolutionary journey through the HR diagram. The labeled positions are characterized in Table 3.5. Dashed lines indicate rapid evolutionary phases

and if, at any time there is a change in the energy generation rate so the temperature gradient must also change in sympathy. Concomitant to this, so, too, must the pressure gradient change. The star accordingly rearranges its internal temperature, pressure and mass distribution until it finds a new stable radius at which gravity is once more held in check. If the surface radius and luminosity of a star change (and the latter is an expression, recall, of the total amount of energy generated per second in the central core), the Stefan-Boltzmann law (Eq. A.4 in Appendix 1 in this book) requires that the temperature of the star must also change. The evolution of a star, as it ages, can accordingly be seen via its motion within the HR diagram – the diagram in which the luminosity of a star is plotted against its temperature (recall Appendix 1 in this book).

The evolutionary track of the Sun in the HR diagram, from its formation to its initial white dwarf phase, is illustrated in Fig. 3.10,

TABLE 3.5 Key ages in the evolution of the Sun. The columns indicate the age in billions of years, the mass, radius, and luminosity in present-day solar units, the temperature in Kelvin and the epoch name abbreviation. The table data is from Schröder and Smith. (*) **Key**: ZAMS = zero age main sequence; *MS* main sequence, *EMS* end main sequence, *RGBT* red giant branch tip, *HB* horizontal branch, *AGBT* asymptotic giant branch tip, *TP* thermal pulsing (Mira variable phase), *PN* planetary nebula development phase, *NWD* onset of the new white dwarf cooling phase

Age (Gy)	$M_\odot$	$R_\odot$	$L_\odot$	T (K)	(*)
0.00	1.0	0.89	0.70	5,596	ZAMS
4.56	1.0	1.00	1.00	5,774	Now
7.13	1.0	1.11	1.26	5,820	MS
10.00	1.0	1.37	1.84	5,751	EMS
12.17	0.67	256	2,730	2,602	RGBT
12.17	0.67	11.2	53.7	4,667	HB
12.30	0.55	149	2,090	3,200	AGBT
12.30	0.55	179	4,170	3,467	TP
>12.3	0.54	0.01	10	50,000	NWD

K-P. Schröder and R. C. Smith, "Distant future Sun and Earth revisited" (available at arxiv.org/pdf/0801.4031v1.pdf

and various key evolutionary moments are described in Table 3.5. Having first initiated hydrogen fusion reactions the new Sun settles onto its zero age main sequence (ZAMS) position. At this zero time it is less luminous, smaller and lower in temperature than the Sun is now when it is 4.56 billion years old. On the main sequence (MS) the Sun steadily generates energy by converting hydrogen into helium within its central core, and this slow composition change results in the Sun becoming more luminous, larger and hotter. After 10 billion years, the hydrogen within the inner 10 % of the Sun has all been converted into helium, and the end main sequence (EMS) stage is achieved. At this time the Sun is 84 % more luminous, 37 % larger, but only 3 % hotter than at the present time.

With the exhaustion of hydrogen in its central core, the Sun has hit an energy crunch. It must now rearrange its internal structure so as to configure a hotter and denser core in order that helium fusion reactions can begin. To do this the inert helium rich core contracts through gravity. At the same time as the core contracts a thin hydrogen burning shell develops around it, and the star's outer radius begins to swell. The gigantism continues, with the star moving upwards along the so-called giant branch, which is characterized by a low surface temperature, large radius and increasing luminosity. At the red giant branch tip (RGBT) the Sun will have swelled to 253 times its current size, and it will be some 2,730 times more luminosity. Its surface temperature, however, will be about half that of present. It is at the RGBT that the shrinking core attains a density and temperature at which the triple-alpha reaction can commence, and in a bright flash the inert core will come to life, vigorously converting three helium nuclei into carbon and energy in the form of a gamma ray photon: $3 \times {}^{4}\mathrm{He} \Rightarrow {}^{12}\mathrm{C} + \mathrm{energy}$. At this time the Sun will settle into the relatively long-lived horizontal branch (HB) stage of helium burning.

In the later part of the RGBT phase the Sun will begin to lose mass from its outer envelop – this material being returned to the interstellar medium. Indeed, between the EMS and HB phase the Sun will shed over 30 % of its mass (a mass equivalent to about 316 Jupiter-like planets). On the HB, the Sun will be about 50 times more luminous, 11 times larger, but about 1,000 K cooler than at present.

The inevitable Malthusian effect, however, soon comes into play. As the helium is steadily consumed within the now increasingly carbon-rich core a new hydrogen burning shell develops, and the Sun will once again be on the move in the HR diagram. This time the Sun will begin to ascend the so-called asymptotic giant branch (AGB), growing in size and luminosity. At the top of the AGB the Sun will be about 150 times larger and 2,100 times more luminous than at present. The life of the Sun, as a star, is now rapidly coming to a close. Having reached the AGBT the Sun's carbon-rich core will become inert, but two shells, one converting helium into carbon and the other, further out, converting hydrogen into helium will develop. This situation literally destabilizes the giant Sun, and it enters into a thermal pulsing (TP) phase. In the TP phase the Sun will vary in both luminosity and size, figuratively wheezing out its last breaths, and it will shed even more mass, ultimately thinning down to about half of its present constitutional bulk. At this time the Sun will once again be bloated in size, swelling to some 180 times larger than at present, and it will again be highly luminous, attaining a luminosity over 4,000 times larger than that on the main sequence.

With the thermal pulsing stage, the end of the Sun will finally have arrived. Its journey in the HR diagram will now be rapid, and at a nearly constant luminosity it will shrink dramatically in size, to become a carbon-rich white dwarf.

As a newly formed white dwarf (NWD) the Sun will be about ten times more luminous than at present, but weigh in at about half of its current mass. The Sun's diminutive size is now comparable to that of Earth (being just 1/100th of its present-day value), although its surface temperature will be some 50,000° – making it 8 ½ times hotter than the Sun on the ZAMS.

The incredibly high temperature of the proto-white dwarf Sun will result in the generation of copious numbers of ultraviolet photons, and these will ionize the surrounding interstellar environment. The Sun will now enter a short-lived, lasting perhaps 15,000–20,000 years, planetary nebula (PN) phase. The high luminosity along with the expansion and contraction cycle that ensues during the AGBT to TP phases will have resulted, recall, in the outer envelope of the Sun being lost into space. The speed of the ejected matter will vary, and faster moving material will plow into slower moving material ahead of it, and a giant bubble of gas around the central white dwarf core will gradually be built up.

The nuclei of the atoms in the inner part of the surrounding bubble will be ionized by the intense flux of UV photons generated by the hot white dwarf; further out, however, where the UV flux is smaller, the ionized nuclei can re-capture their wayward electrons and thereby produce emission line radiation. Since the density of material in the nebula bubble will be so low that otherwise unexpected emission lines can be generated from the oxygen and nitrogen atoms. These atoms are formed by fusion reactions in the deep interior of the star and then dredged up into the outer envelope by deep convection currents. The so-called forbidden lines produced by oxygen have a distinctive green color, while those of nitrogen are red. The bubble, growing to perhaps a light year across in physical size, literally begins to shine (Fig. 3.11).

The evolution of α Cen A and B will be contrasted against that of the Sun in the section below. In the meantime, however, the catastrophic influences of an aging Sun on the rest of the Solar System and specifically its habitability zone will be considered.

FIG. 3.11 Planetary nebula Abell 39. This beautifully symmetric nebula is located about 6,800 light years from the Solar System and is some 5 light years across. The parent white dwarf is clearly visible in the image and is located slightly off center. Both the mass contained within the nebula as well as the mass of the white dwarf are estimated to be about 0.6 solar masses, and to attain its present size the nebula must have been expanding for approximately 22,000 years (Image courtesy of WIYN/NOAO/NSF)

3.12 The End

The evolutionary changes that will take place in the Sun will have three direct effects on the rest of the Solar System. First, the Sun's increasing luminosity will drive planetary temperatures higher; second its radius will expand to beyond the orbits presently occupied by the inner most terrestrial planets; and third, its loss of mass will cause all planetary orbits to change. All three of these effects will combine to destroy any and all of the life-forms that presently exist within the Solar System's habitability zone.

The temperature of a planet is primarily determined by the size of its orbit and the Sun's luminosity and through a dynamical balance between the Sun's insolation and the energy radiated back into space by the planet. The planet can be treated as an approximate blackbody radiator (see Appendix 1 in this book), and accordingly, the equilibrium temperature is determined by the relationship:

$$T = 278\left(\frac{(1-A)L}{\varepsilon d^2}\right)^{1/4} \tag{3.2}$$

where L is the Sun's luminosity (in units of $L_\odot$), d is the orbital radius in astronomical units, T is the planet's temperature, A is the albedo and ε is the emissivity of the planet. The latter two parameters account for the fact that no planet is a perfect blackbody radiator. The albedo term accounts for the fact that not all of the Sun's energy is absorbed at the planet's surface (some is directly absorbed by atmospheric gases, while the rest is simply reflected back into space), while the emissivity term accounts for the fact that, once heated, not all of the absorbed energy is radiated straight back into space (some of it is conducted into the ground below the surface).

In deriving Eq. 3.2 it is assumed that the planet has a sufficiently rapid period of rotation that its entire surface is uniformly heated and radiates energy back into space. In the slower rotation case, where only one hemisphere is heated and radiates energy back into space, the 278 numerical term is multiplied by a factor of $2^{1/4} \approx 1.2$. Additionally, it is worth noting that the temperature of a planet does not depend on its physical size but only on its orbital distance from the Sun, along with its albedo and emissivity characteristics. For Earth, typical albedo and emissivity average values are $A = 0.3$, and $\varepsilon = 0.6$; for the Moon they are typically taken as $A = 0.1$ and $\varepsilon = 0.9$.

Substituting values appropriate to Earth at the present time, with $L = 1\ L_\odot$, Eq. 3.2 indicates that $T(\text{Earth}) = 288$ K, which is indeed the characteristic global average temperature for Earth. The result just derived is partly a cheat in that Eq. 3.2, through the introduction of the albedo and emissivity terms, has been adjusted to include a greenhouse heating effect. This effect relates to the manner in which the atmosphere modifies the surface re-radiation process. The solar energy that initially warms the surface of Earth penetrates the atmosphere at wavelengths corresponding to visible light.

The energy re-radiated into space by the warmed surface, however, is emitted predominantly as longer wavelength infrared radiation. Molecules such as water vapor H_2O, carbon dioxide CO_2 and methane CH_4, which reside within the atmosphere, however, readily absorb this longer wavelength radiation. Accordingly, the atmosphere itself is warmed and then back-heats the surface of Earth. In essence the atmosphere acts like a blanket, keeping Earth's surface warmer than it would otherwise be without an atmosphere containing greenhouse gases. If Earth could radiate its excess heat energy back into space without hindrance, then the emissivity would be $\varepsilon = 1$, and Eq. 3.2 indicates that an equilibrium temperature of $T = 254$ K would result. The additional greenhouse heating effect, therefore, amounts to adding an additional 34° to the global average temperature at the present time.

In terms of habitability, life on Earth is only possible because some greenhouse heating does take place. Indeed, the greenhouse heating of Earth's atmosphere currently keeps the average global temperature at a comfortable 15 °C, rather than the ocean-freezing, life arresting −19 °C that it would otherwise have. As with all things, however, moderation is the key. Too little greenhouse warming and the oceans freeze; too much greenhouse warming and the oceans literally evaporate away. Although human agriculture, oil dependency, and industrial activity are currently driving a potentially disastrous near-term climate-altering greenhouse warming, the Sun's increasing luminosity will ultimately tip the balance towards the utter destruction of life on Earth.

The characteristic variation of the Sun's luminosity, as it ages on the main sequence, is described by the relationship[56]

$$L(t) \approx 0.8\left(1 - 0.24\frac{t}{\tau}\right)^{-15/17} \quad (3.3)$$

[56] This formula is based on an analytic model development; see, for example, Hansen and Kawaler's *Stellar Interiors* (Note 55 above). The odd-looking 15/17 power comes about because of the adoption of a specific, so-called Kramer's approximation opacity law.

where the time t is expressed in billions of years, $\tau = 4.5$ Gyr is the present age of the Sun and where $L(t)$ is given in solar units.

At the present epoch, therefore, the Sun's luminosity is increasing by about 1 % every 150 million years. In one billion years from now, its luminosity will be 7 % greater than at present; two billion years hence it will be 14 % more luminous. For a 10 % increase in the Sun's luminosity, Earth's radiative temperature, as described by Eq. 3.2 with $\varepsilon = 1$, will increase by nearly 7 K. This at first seems like a small increase, but it is a small change that sets many larger forces into catastrophic motion – indeed, the destruction of Earth's biosphere has irreversibly begun. With the ongoing and unstoppable increase in Earth's surface temperature it is the sleeping giants of the oceans and atmosphere that first begin to suffer.

The first of the destructive positive feedback mechanisms that comes into play is that between the atmosphere and the ground. Specifically it is the rate at which silicate rocks are weathered away that becomes increasingly important. One of the key weathering reactions runs according to this sequence: $CaSiO_3 + CO_2 \Rightarrow CaCO_3 + SiO_2$, with the end result that the original atmospheric carbon along with its two oxygen companions are captured into solid carbonate and silicate phases.

As the temperature increases so the weathering process runs more rapidly and draws down more and more atmospheric CO_2. On the one hand this is a good thing, since it removes an important greenhouse gas from the atmosphere, and the CO_2 reduction will partly offset the temperature increase that Earth would have otherwise suffered as a consequence of the Sun's steadily increasing luminosity. On the negative side, however, plants require CO_2 to undergo photosynthesis, and once the CO_2 level in the atmosphere drops below 150 parts per million (ppm)[57] then many plant species will begin to die out. Once the atmospheric CO_2 level falls below 10 ppm, all photosynthesizing plant life will come to an end. With the suffocation of plant life, the animal food chain will begin to collapse, and soon all megaflora and fauna will have disappeared, leaving only microbial life to claim Earth as a home.

[57] The current atmospheric concentration of CO_2 is 400 ppm – and due to human activity it is rising rapidly!

Although the enhanced weathering rate reduction in atmospheric CO_2 will moderate the rate at which Earth warms, in the future it cannot fully reverse the heating trend. It is now that a second ultimately more destructive positive feedback mechanism will begin to develop. As the temperature rises, so too will the rate at which ocean evaporation takes place. Since water vapor is one of the most potent of greenhouse gases, the larger its concentration in Earth's atmosphere the higher the greenhouse atmospheric heating effect becomes, and this causes an even higher rate of ocean evaporation. This is the runaway moist greenhouse effect. Indeed, the complete evaporation of all Earth's oceans come about relatively swiftly and proceed at an exponential rate. In short order Earth will effectively lose its ability to support any plant, fish or animal life. The biosphere as we presently know it will not only be dead, it will be irrevocably dead, and no subsequent actions can bring it back to life. Indeed, the oceans themselves are essentially lost into space as a result of the photo-disintegration of the water molecule into its constituent parts, two hydrogen atoms and one of oxygen, with the lightweight hydrogen atoms being rapidly lost through Earth's exosphere.

What might remain alive at this stage is an open question, but it is easily envisioned that variously adapted microbes and extremophiles, buried deep below Earth's surface, might eke out a continued existence for a few more billions of years. But even they are ultimately doomed (as we shall see).

Adding insult to injury the demise of life on Earth will potentially add a distinctive chemical marker to the atmosphere – one that will be broadcast to any extraterrestrial astronomer that "life once thrived here." The chemical signal will come about through the decay of dead bodies. Indeed, Jack O'Malley-James (St. Andrews University) and co-workers modeled[58] the gases that would be produced during a rapid mass extinction event, and, it turns out, the mass die-off produces a significant quantity of the gas methanethiol (CH_3SH). While methanethiol has a relatively short atmospheric lifetime of about 350 years, it eventually disassociates to

[58]Jack O'Malley-James (St. Andrews University) and co-workers, "Swansong Biospheres II: the final signs of life on terrestrial planets near the end of their habitable lifetimes" (article available at arxiv.org/pdf/1310.4841v1.pdf).

form ethane (C_2H_6), which has a much longer atmospheric survival time. Turning this scenario on its head, O'Malley-James et al. note that should we find an exoplanet having an atmosphere revealing a signature of methanethiol or one rich in ethane, then this would indicate the discovery of a planet that has undergone a recent catastrophic die-off. The irony of such a discovery, of course, would be that through finding the signature of mass death we would discover that life had evolved somewhere else in the galaxy.

Many complex, large-scale and competing processes will come into play in the destruction of Earth's biosphere, but the general consensus, from multiple approaches of analysis, suggests that the CO_2 starvation of plant life will take place in about 1–1.5 billion years from the present. The moist greenhouse loss of the oceans will be in full swing a few hundred million years later. In short, 2–2.5 billion years from now Earth's surface will constitute a lifeless, scorched desert. But the devastation is not just limited to Earth. The Sun's lethal bite will extends to the entire Solar System.

At present it is not clear that Earth is the only body within the Solar System that supports, or could have supported, life in the past. There is a good possibility that underground bacterial life thrives to this very day on Mars, and there is also a very good chance that bacterial life could exist in the under-ice ocean of Jupiter's moon Europa (and possibly in the sub-surface liquid regions of Saturn's moon Enceladus). Bacterial life may also survive in other Solar System locations not even thought of and/or investigated yet. Indeed, if life on Earth tells us anything, it tells us that life is vigorously opportunistic and highly tenacious.

The possible life reservoirs of Mars and Europa, however, are not safe from the increasing Sun's luminosity. As Table 3.5 indicates once the Sun begins to climb the red giant branch its luminosity increases dramatically. Indeed, at the tip of the red giant branch (RGBT) it will briefly be over 2,000 times more luminous than at present. What all this means for the Solar System's habitability zone is that it will be pushed further and further outwards into the deeper Solar System, sterilizing the entire region between its innermost edge and the Sun as it goes. Even Moon life, such as that which might exist in Europa's ocean, will be killed off at this stage. Indeed, Europa itself will only just survive. At the orbit of

Jupiter, $d = 5.2$ AU, Eq. 3.2 indicates that the present (perfect blackbody radiator) temperature is 122 K. The water ice at the surface of Europa will begin to sublimate once the temperature there exceeds 273 K, and this will occur when the Sun's luminosity is greater than 85 $L_{\odot}$ (assuming an albedo of $A = 0.7$ for Europa). The Sun will pass this luminosity threshold in about 7.5 billion years from the present, and within a few millions years of time thereafter all the ice and oceans of Europa will be gone, leaving just a small, lifeless rocky core to orbit around Jupiter.

Ultimately all natural life, along with many of its habitats in the Solar System, will be destroyed by the Sun. The more immediate consequences of global warming, the drawing down of atmospheric CO_2 by weathering and the resultant destruction of photosynthesizing plant life, as well as the runaway moist greenhouse loss of the oceans, might possibly be mitigated against through geo- and astro-engineering options. With respect to the latter, the author has previously discussed, in the book *Rejuvenating the Sun and Avoiding other Global Catastrophes* (Springer, 2008), some of the large-scale engineering projects that our distant descendants might one day initiate in order to modify the Sun and possibly save terrestrial life. There are survival options open to our descendants, but none will stay the Sun's ultimate hand of execution – indeed, nothing (not even the Solar System) lasts forever.[59]

Master science fiction writer and social commentator H. G. Wells described the end of life on Planet Earth in his 1895 book *The Time Machine*. The scene is bleak and "the huge hull of the Sun" overlooks, in brooding manner, the demise of all animate matter. Although Wells worked within the realms of then known astronomy, his descriptions of the demise of life on Earth failed (since the effect was then unknown[60]) to take into account the loss of the oceans. Additionally, though Wells correctly alludes to the expansion of planetary orbits, he grossly overestimates the effect, and at the same time fails to account for the red giant expansion of

[59] Of course, forever is a long time. The ultimate end of matter essentially depends on the stability of the proton. Current estimates for the proton decay time suggest that it cannot be less than 10^{32} years.

[60] For details on the greenhouse effect and atmospheric heating, see, for example, the author's book, *Terraforming: The Creating of Habitable Worlds* (Springer New York, 2009).

the Sun. Indeed, Wells dramatically writes of a solar eclipse being observed from the dying Earth and attributes it to the passage of Mercury between the Sun and Earth. As we shall see, it is true that planetary orbits will expand as the Sun ages, but the expansion is far too small to save Mercury and most probably Venus from utter destruction. These two innermost terrestrial planets are doomed to a fiery death in the Sun's bloated red giant atmosphere. Remarkably, a preview of this future planetary annihilation is in the process of being played out around the distant Sun-like star Kepler-56. Located nearly 1,700 light years away, this star, it is estimated by Daniel Huber (NASA Ames Research Center) and co-workers, is already in its end main sequence phase and in some 150 million years from the present it will begin to swell into a red giant. When this red giant expansion phase begins, two of its three known planetary companions (Kepler-56b, a Neptune-like planet, and Kepler-56c, a Saturn-like planet) will be consumed within its billowing envelope.

Why will planetary orbits expand as the Sun evolves? The answer relates to the mass loss that occurs as the Sun first ascends the giant branch (GB in Fig. 3.10) and then later the asymptotic giant branch (AGB in Fig. 3.10). Indeed, over these two evolutionary phases the Sun will lose nearly 50 % of its present mass (see Table 3.5). In direct response to this mass reduction the orbits of all the planets within the Solar System move outwards in order to conserve their orbital angular momentum. This condition holds true provided the timescale for mass loss from the Sun is long compared to the orbital period of the planet – a condition that is in fact well satisfied. Accordingly the quantity $M(t)\ a(t)$ must be constant at all times t, where $M(t)$ is the mass of the Sun, and $a(t)$ is the planet's orbital radius. The orbital radius at time t will accordingly be $a(t) = a(0)\ M(0)/M(t)$, where the $t = 0$ terms can be taken as the present-day solar mass and planetary orbital radius. As shown in Table 3.5, the Sun loses about a third of its mass as it moves away from the main sequence towards the tip of the red giant branch. Accordingly the orbital radii of each planet will increase by a factor of 3/2. For the planets Mercury and Venus we currently have $a(0) = 0.4$ au and 0.72 au, respectively. At the red giant tip

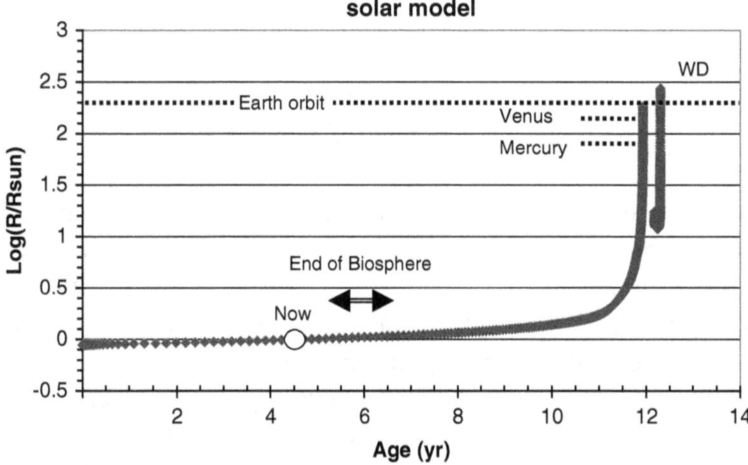

FIG. 3.12 The time variation in the Sun's radius. The open circle indicates the present radius and age of the Sun, and the approximate times in which the biosphere will be destroyed and the oceans lost through the runaway moist greenhouse effect are indicated. The Sun will begin to swell dramatically in about five billion years, and it will be at this time that the orbits of Mercury and Venus will be overrun by its expanding red giant envelope. The dotted horizontal line indicates the present radius of Earth's orbit. The Sun ultimately evolves into a white dwarf (WD) having a radius about 100 times smaller than at present. The solar model has been derived from the EZ-web server developed by Richard Townsend (www.astro.wise.edu/~townsend/)

these orbital radii will have increased to 0.6 and 1.1 au respectively. Table 3.5 indicates, however, that at the red giant tip the Sun will have expanded to a radius 256 times larger than at present – this radius corresponds to 1.2 au. Here then is the crunch – the Sun will swell outwards to engulf the orbits of Mercury and Venus (Fig. 3.12). Earth's orbital radius will have increased to 1.5 au (322 $R_\odot$) at this time, and though it will undergo a considerable roasting it will probably survive against complete physical destruction.[61] The amount of surface mantle lost by Earth during the RGBT stage can be estimated from the sublimation rate of surface material. In

[61] The term probably is used since one of the key factors in determining Earth's survival will be the extent of its tidal interaction with the red giant Sun. The complexities of such calculations are discussed by Schröder and Smith (Note 44 above), and by Rasio et al. in their research paper "Tidal decay of close planetary orbits" (*Astrophysical Journal,* **470**, 1187, 1996).

this manner, the rate of change in Earth's physical radius dR/dt is given by

$$\frac{dR}{dR} = \frac{1}{\rho}\frac{d\sigma}{dt} = \frac{1}{\rho}\sigma_0\sqrt{\frac{T_0}{T}}\exp(-T_0/T) \tag{3.4}$$

where $d\sigma/dt$ is the mass sublimation rate of surface material per unit area and ρ is the material density. In Eq. 3.4 it has been assumed that the primary surface material is pure olivine, and accordingly the mass sublimation formula deduced Hiroshi Kimura (Kobe University, Japan) and co-workers has been used.[62] Earth's temperature during the Sun's red giant tip (RGT) phase is derived from Eq. 3.2, assuming $A = 0$, $\varepsilon = 1$, $L = 2730\ L_{\odot}$ (see Table 3.5) and $d = 1.5$ au. Accordingly, Earth's temperature at this stage is some 1,650 K. Substitution of numbers into Eq. 3.4 now yields a radius reduction rate of $dR/dt \approx 2 \times 10^{-10}$ m/s. Given a typical phase time of $\tau \approx 100{,}000$ years, Earth's radius is uniformly reduced by about $\tau\ (dR/dt) \approx 0.6$ km.

During the Sun's asymptotic giant branch phase its luminosity increases to some 3,000 $L_{\odot}$, but Earth's orbit has expanded to 1.8 au. Accordingly Earth's radius will be reduced by an additional 0.1 km. Although Earth will probably survive as a physical entity as the Sun ages, its radius will be reduced by about 1 km. This reduction in the radius, while insignificant to Earth itself, is exactly the layer in which the post ocean-loss bacterial life resides. As the Sun moves into its thermal pulsing and planetary nebula phases its innermost planet will be a silent, uninhabited, scorched and sterile world.

Upon having its orbit overrun by the outer atmosphere of a star, a planet's demise will follow in short order. The approximate timescale of destruction will be of order $\tau = M_P/M_{acc}$, where M_P is the planet mass and M_{acc} is the rate at which the planet is accreting material from the star. The argument here is that the planet will be destroyed in about the same amount of time that it takes to encounter its own mass in the form of atmospheric gas. The accretion rate to a first approximation is related to the planet's

[62] The sublimation formula is taken from the research paper by Hiroshi Kimura et al., "Dust grains in the comae and tails of sungrazing comets: modeling of their mineralogy and morphological properties" (*Icarus*, **159**, 529, 2002).

cross-section area, and accordingly, $M_{acc} = \pi R_P{}^2 \rho_{atm} V_P$, where R_P is the planet's radius, V_P the planet's orbital velocity and ρ_{atm} is the density of the star's outer envelope. Characteristic numbers for Mercury and Venus indicate orbital decay and destruction lifetimes of 117 and 384 years, respectively.[63]

The energy released by the accretion and destruction of a planet is essentially that of its orbital kinetic energy Ke, and the characteristic luminosity of such events will be $L_{acc} = \mathrm{Ke}/\tau$, where τ is the destruction timescale. For Mercury and Venus the additional energy per second added during their destruction and accretion phase is of order 1/1,000th that of the Sun's present luminosity. By analogy, therefore, the demise of terrestrial exoplanets in orbit around other stars is something that would be very difficult to observe – even if one knew where to look. Kepler-56 is the current future exception to this statement. Indeed, with such short timescales an observer would have to be extremely lucky to catch a planet destruction event in progress.

Remarkably, however, this may in fact have actually happened in the case of the odd sudden brightening of V838 Monocerotis. Although alternative models have been proposed for the observed, 80-day-long, enhanced-luminosity outburst from this particular star, Alon Retter (Pen-State University) and colleges have suggested that it might have been caused by the accretion and destruction of three Jovian planets.[64] A Jupiter-mass planet orbiting a Sun-like star at an orbit distance of 1 au will have an orbital decay time of about 1,250 years and would contribute an additional 0.1 $L_{\odot}$ in additional energy output to the star over this time interval. If the accretion time is much shorter, however, as

[63] The calculation assumes $\rho_{atm} = 10^{-4}$ kg/m^3, which is taken from the numerical models developed by William Rose and Richard Smith in their research paper, "Final Evolution of Low-Mass Star II" (*Astrophysical Journal*, **173**, 385, 1972).

[64] A planetary accretion hypothesis has been developed by Alon Retter and co-workers to explain the unprecedented outburst from a star called V838 Monocerotis. The details of their calculations are given in the research article, "The planets capture model of V838 Monocerotis: conclusions for the penetration depth of the planet/s" (available at arxiv.org/pdf/astro-ph/0605552v2.pdf). Since three peaks were recorded in the outburst light curve of V838 Monocerotis, the accretion of three Jovian planets was implied. Given the detection of many hot Jupiter exoplanets orbiting close in to their parent stars, it is likely that outbursts similar to that of V838 Monocerotis should be fairly common. Retter and co-workers suggest that one such event per year per galaxy might be expected.

would be the case for exoplanet hot Jupiters that have orbital radii of just a few tens of solar radii, then an even greater luminosity increment would be realized. Retter and co-workers, in fact, argue for destruction times of order of tens of days in the case of V838 Monocerotis, and so the resultant luminosity contribution will be of order 10^3 $L_\odot$.

In principle, the destruction rate of terrestrial planets within the galaxy can be gauged by measuring the formation rate of white dwarfs – the latter objects being directly associated with those systems that have undergone their red giant and asymptotic giant branch evolution (the phase at which terrestrial planet destruction is going to take place). In a detailed study of white dwarf statistics, undertaken by James Liebert (University of Arizona) and co-workers in 2005,[65] it was found that the formation rate of such objects amounts to about 10^{-12} per cubic parsec of space per year. Taking the volume of the galactic disk to be of order 4×10^{11} pc^3 (based upon a radius of 16 kpc and a disk height of 0.5 kpc), the number of white dwarfs formed in the galaxy is of order 0.4 per year.

Most of the progenitor stars forming white dwarfs will be in the spectral type range F, G and K, and Wesley Taub (Jet Propulsion Laboratory) has recently estimated, via Kepler spacecraft data, that 30 % of such stars will have planets in their habitability zones. On this basis the number of terrestrial planet deaths within the entire galaxy, that is, the number of planetary systems being overwhelmed by their expanding red giant radius of the parent stars, will be of order $0.4 \times 0.3 = 0.12$ per year. In other words we might expect that once every 8–10 years, somewhere in the Milky Way Galaxy, one or more terrestrial planets will be destroyed as their parent star evolves away from the main sequence.

3.13 The End: Take Two

Although the Sun and the α Centauri system will no longer be physically close several billion years from now, they will nonetheless both begin to destroy their respective habitability zones at

[65] See the research paper by James Liebert, P. Bergeron and J. Holberg, "The formation rate and mass and luminosity functions of DA white dwarfs from the Palomar Green Survey" (*Astrophysical Journal* supplement series, **156**, 47, 2005).

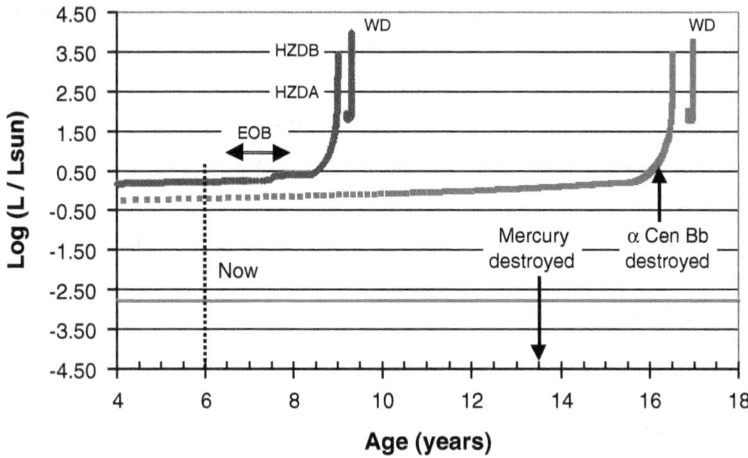

FIG. 3.13 Luminosity versus time diagram for the α Cen A (*top, upper left* sequence), α Cen B (*middle*, and *upper right* sequence), and Proxima (*lower, solid line*). The time at which Earth's biosphere will be destroyed is indicated by EOB, while the destruction of the habitability zones around α Cen A and α Cen B are indicated by HZDA and HZDB, respectively. The time at which Mercury and Venus will be destroyed in the red giant Sun's envelope is indicated, along with the time at which α Cen Bb will meet its demise in the expanding envelope of α Cen B. The EZ-web server developed by Richard Townsend (www.astro.wise.edu/~townsend/) has been used to obtain the evolutionary model data for α Cen A and B (based upon Table 2.2 characteristics)

about the same time. Fig. 3.13 shows the luminosity versus time diagrams for α Cen A, α Cen B and Proxima. Since the α Centauri system is about 1.5 billion years older than the Solar System, and because α Cen A is more massive than the Sun, it will become a red giant in about three billion years from the present. At this time the habitability zone around α Cen A will be destroyed. We don't presently know, of course, if there are any habitable planets within the Centauri system, but we do know where any such planets must reside, and this enables us to calculate their future demise just as we did with the Sun and Solar System earlier. Though purely a coincidence, the destruction of the habitability zones in the Solar System as well as those around α Cen A and B (see below for details) will occur at about the same time.

As described earlier, the planet stability zone for α Cen A stretches out to a distance of about 4 au. Its habitability zone is situated between 1.17 and 2.23 au (recall Fig. 2.27). The detailed numerical models indicate that these radii will be overrun by the expanding red giant envelope of α Cen A in approximately three billion years from the present time. Any Earth-like planets (that is, having an Earth-like atmosphere and ocean system) within the habitability zone around α Cen A will begin to undergo a runaway moist greenhouse loss of their seas in about two billion years from the present – indeed, at about the same time as our Earth will lose its aqueous biosystem.

The stable planetary zone around α Cen B is of comparable size to that of α Cen A, although its habitability zone resides in the region between 0.56 and 1.10 au. It is the smaller mass and concomitant lower main sequence luminosity of α Cen B that brings its habitability zone closer in than that of α Cen A. Since α Cen A is more massive and much more luminous than α Cen B (recall Table 2.2), it will have a distinct effect upon the equilibrium temperatures of any planets within the habitability zone of α Cen B. Indeed, if, for example, there is a habitable planet at, say, 0.8 au from α Cen B, then the energy flux that it receives from both stars will be

$$T = 278\sqrt[4]{\left(\frac{L_B}{0.64} + \frac{L_A}{ab}\right)} \qquad (3.5)$$

where a and b are the orbital semi-major and semi-minor axis, expressed in astronomical units, of the α Cen AB binary (see Appendix 3 in this book) and the luminosity terms are expressed in solar units. In deriving Eq. 3.5 we have averaged the flux received from α Cen A by our hypothetical planet in orbit around α Cen B over the entire orbital period of the system. From Eq. 3.5, the temperature of our imagined planet due to α Cen B alone is 261.4 K. This temperature is increased to 261.6 K when the additional present-day flux from α Cen A is included. When at their closest approach distance to each other the presence of α Cen A will increase the temperature of our hypothetical planet by about 1 K. At this time, therefore, the existence of α Cen A has but a small effect upon the habitability zone characteristics of α Cen B.

As α Cen A evolves towards its bloated red giant stage, however, the situation changes dramatically. Indeed, as it evolves through the red giant tip and into the asymptotic giant branch phase, α Cen A will add an additional 25 and 50 K to the equilibrium temperature of any planet located within the erstwhile habitability zone of α Cen B. Indeed, like that around the Sun and α Cen A the habitability zone of α Cen B will be no more within four billion years from the present. The physical demise of α Cen Bb will not come about until α Cen B itself evolves away from the main sequence. As indicated in Fig. 3.13, the fiery consumption of α Cen Bb will occur in about ten billion years from the present – at a time several billion years after the destruction of Mercury and Venus within the glowing outer envelope of the red giant Sun.

Almost going unnoticed in Fig. 3.13 is the horizontal line describing the time evolution in the luminosity of Proxima Centauri. The hyper-slow evolution of Proxima will be discussed shortly; for the next several tens of billions of years, however, it will hardly change its energy output at all, and accordingly it will provide a very stable environment, energetic flares aside, for any planets located within its habitability zone. Rather than having a lifetime measured in ten or so giga (10^9) years, as in the case of the Sun and α Cen A and B, Proxima evolves on a timescale of tera (10^{12}) years. In about 4 tera years time Proxima will attain its maximum luminosity of about 0.025 $L_\odot$ – making it some 15 times more luminous than at present ($L_{now} = 0.0017\ L_\odot$; recall Table 2.2). This increase in luminosity will push the innermost edge of Proxima's habitability zone from its current 0.023 to 0.088 au.

Since there is no limit to the stability zone for any planets orbiting Proxima, the outward expansion of the habitability zone may not be critical to the survivability to any life forms that may have evolved in or colonized the system. The energy flux contribution from α Cen A is entirely negligible at the present time with respect to the heating of any planets that might exist in orbit around Proxima. Even in its later stages, when α Cen A is a thermally pulsing Mira variable-like star, with a luminosity of order $10^4\ L_\odot$, its heating effects will be minimal, less than 0.5 K, in fact, for a Proxima orbiting planet, provided Proxima comes no closer than 2,000 au to α Cen AB. At present, of course, we do not know the exact orbit for Proxima, but by the time that α Cen A is approaching its final evolutionary phases towards becoming a white dwarf, three billion years from now, Proxima will have fled its α Centauri natal nest anyway. At this time, freed from its gravitational thralldom, Proxima will set flight on its own course around the galactic center. Indeed, as the unstoppable march of time takes us through the next ten billion years, α Centauri will gradually, step by step, begin to break apart.

3.14 The Dissolution of α Centauri

As the Sun slims down in mass during its red giant and asymptotic giant branch phases, the orbits of the planets begin to move outward towards larger radii. This outward migration of the planets, as we have seen, is too small to save Mercury and Venus from a fiery demise in the outer envelope of the expanding red giant Sun, but it is possibly enough to save Earth from physical destruction. Given that mass loss from the Sun results in the radial expansion of planetary orbits, what, we may ask, happens to a binary star system as each of its components age and lose mass into interstellar space?

The two-body problem in which isotropic mass loss occurs has been studied since the late nineteenth century and is often discussed under the guise of the Gylden-Meshcherskii[66] problem. It was Swedish astronomer Johan Glyden who first developed the equations to study in 1884, while Russian mathematician Ivan Meshcherskii showed in 1893 that under a set of special conditions analytic solutions to Gylden's equations are possible. The differential equations developed by Gylden describe the secular variations in the orbital eccentricity e and semi-major axis a in terms of the mass loss rate $d\mu/dt$, where $\mu(t)$ is the total system mass at time t. Specifically, the equations to solve for are

$$\frac{da}{dt} = -\frac{a(1+ecosE)}{1-ecosE}\frac{1}{\mu}\frac{d\mu}{dt} \quad (3.6)$$

[66] We note here that there are many variant spellings of Meshcherskii in usage.

for the time variation in the orbital semi-major axis,[67] and

$$\frac{de}{dt} = -\frac{(1-e^2)cosE}{1-ecosE}\frac{1}{\mu}\frac{d\mu}{dt} \qquad (3.7)$$

for the time variation in the orbital eccentricity. The E term appearing in our two equations is the orbital eccentric anomaly (see Appendix 3 in this book) – a quantity that essentially keeps track of the time variation in the moment of closest approach (the periastron). Equations 3.6 and 3.7 can be solved for numerically when the initial orbital parameters are known and when the mass loss rate is determined.[68] These solution conditions are met with respect to the binary pair α Cen A and α Cen B, and will be discussed shortly, but since the exact (or even approximate) orbit of Proxima is not known at the present time a secondary method must be applied to investigate the future of the α Cen AB plus Proxima (α Cen C?) pairing.

To determine the stability of Proxima's orbit the so-called mass-loss index Ψ, as defined by Dimitri Veras (Cambridge University) and co-workers,[69] can be used. This dimensionless constant is expressed in terms of the mass loss timescale and the orbital period P, so we have:

$$\Psi = \frac{\tau_{ML}}{n\mu} = \frac{1}{2\pi}\left(\frac{dM}{dt}\right)\left(\frac{a}{\mu}\right)^{3/2} \qquad (3.8)$$

where, in our case, $\tau_{ML} = M/(dM/dt)$ is the stellar mass loss timescale and where μ is the total system mass of α Cen A and B plus Proxima. Additionally, n and a correspond to the mean orbital motion ($n = P/2\pi$) and orbital semi-major axis of Proxima's orbit. Detailed calculations indicate that once $\Psi \geq 0.5$ then escape from

[67] The full set of time variable equations describing the two-body mass-loss problem are given by Dimitri Veras et al., "The great escape: how exoplanets and smaller bodies desert dying stars" (*Monthly Notices of the Royal Astronomical Society*, **417**, 2104, 2011).

[68] We note here that when the eccentricity is exactly zero – corresponding to a circular orbit. Then we recover from Eqs. 3.6 and 3.7 our earlier approximation for the orbit expansion $a(t) = a(0)\ M(0)/M(t)$.

[69] The mass loss index is defined by Dimitri Veras and Mark Wyatt (both at Cambridge University) in, "The Solar System's post-main sequence escape boundary" (article available at arxive.org/pdf/1201.2412v1.pdf).

the system is inevitable. For escape with $\Psi = 0.5$, we find critical semi-major axis limits of 10^6, 2×10^5 and 4.5×10^4 au as the mass loss for α Cen A increases from 10^{-8}, to 10^{-7} to 10^{-6} solar masses per year respectively.

Recalling the orbital semi-major estimates discussed earlier, we find that for *a* varying between the limits of 8,500 au and 270,000 au Proxima will escape from the fold of the Centauri system once the mass loss rate of α Cen A exceeds 10^{-5} and 5×10^{-8} solar masses per year respectively.[70] Although the former of these mass loss rates will be achieved during the asymptotic giant branch phase, the latter could be achieved earlier during the red giant phase. Either way, within the next 3.5 billion years, Proxima (assuming it does qualify as being α Cen C at this time) will no longer remain gravitationally bound to α Cen AB. The split will be final and irreversible, with Proxima quietly drifting further and further away from α Cen AB into the dark expanse of the galaxy.

In addition to driving Proxima from the family fold, the mass loss experienced by α Cen A as it moves through its post-main sequence evolutionary phases will additionally cause the orbit of α Cen AB to change dramatically. Solving Eqs. 3.6 and 3.7 numerically[71] reveals that the orbital eccentricity increases as more and more system mass is lost. This effect is shown with respect to the closest and greatest separations of α Cen A and B in Fig. 3.14. The initial steep rise in the apastron (greatest separation) begins as α Cen A evolves through its significant mass-losing phases as a red giant. During this time its mass is whittled down from the present 1.105 $M_\odot$ to its final white dwarf mass of 0.67 solar masses. Also during this time interval, the orbital eccentricity of the α Cen AB binary increases from its present value of 0.52–0.94 – pushing it close to the unbound orbit limit of $e \geq 1.0$. The semi-major axis also increases as α Cen A loses mass, and this results in the apastron distance increasing to 276 au – a 25 times increase compared to the current value.

[70] This escape condition is based upon standard Newtonian physics, and the MOND condition discussed in Sect. 2.17, if it applies, would require an entirely different approach to the escape condition problem.

[71] The details are given by the author in the research paper, "The far distant future of alpha Centauri" (*Journal of the British Interplanetary Society*, **64**, 387, 2012).

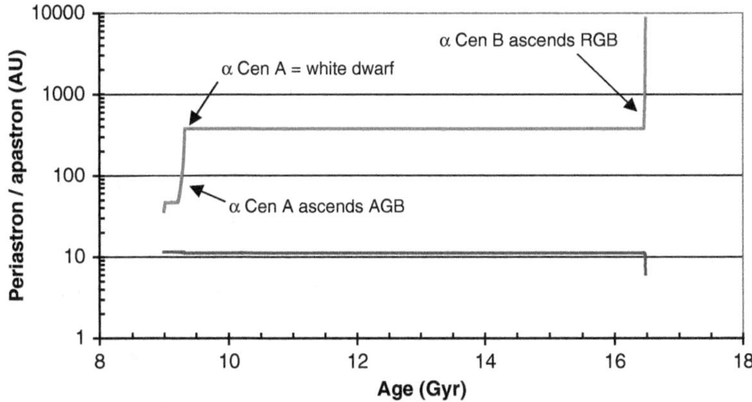

FIG. 3.14 Time variation in the orbital separation – periastron lower line, apastron upper line – of α Cen A and B. The eventual dissolution of the system occurs as α Cen B enters its early red giant phase at a time set some 12 billion years from the present

Once α Cen A has evolved into its initial white dwarf configuration, in about three billion years from the present, α Cen AB will have become a wide binary pairing with an orbital period of 2,136 years. The periastron distance, however, hardly changes during this evolutionary phase, since the increase in the eccentricity is offset by the concomitant increase in the semi-major axis.

Once α Cen A has evolved into its white dwarf configuration, the system's orbital evolution enters a quiescent phase. The orbital eccentricity only begins to change again once α Cen B evolves away from the main sequence and begins to lose mass into space – at a time set at about 12 billion years hence. As α Cen B starts to lose appreciable quantities of mass, the system eccentricity begins to increase again, and some 12.3 billion years from now the system will become unbound. At this time α Cen A (now a cooling and sedentary white dwarf) and α Cen B (now a burgeoning red giant) will set off along independent orbits around the galactic center. The currently familiar triumvirate of α Centauri system will be no more.

3.15 When Proxima Dies

The pace of time now becomes even slower than glacial. For Proxima, change takes eons to accumulate. We shift by a factor of a thousand in timescale; change is measured in trillions rather than billions of years now. The temporal and spatial infinities that so terrified French philosopher Voltaire must now become real before Proxima begins to noticeably change from its present appearance.

At some six billion years old, Proxima has hardly even begun to age. Indeed, it barely qualifies as having entered into Winston Churchill's "end of the beginning" phase. The internal clock that ticks out the evolutionary pace of Proxima runs slowly – incredibly slowly. The main sequence lifetime of star, during which it generates internal energy through hydrogen fusion reactions (recall Fig. 2.3), was described earlier. For Proxima, this time embraces a staggering six trillion years. Although it contains only slightly more than a tenth of the mass of the Sun, the low luminosity of Proxima (Table 2.3) dictates a hydrogen-burning (main sequence) lifetime that is some 500 times longer than that of the Sun. This extended time comes about partly because of Proxima's low luminosity. But mainly because of its interior being fully convective and therefore well mixed, Proxima can access its entire hydrogen fuel supply, rather than just the innermost portion, as in the case of the Sun.

The evolutionary track of Proxima in the HR diagram is shown in Fig. 3.15. This diagram is based upon the detailed numerical calculations presented by Gregory Laughlin (University of California, Santa Cruz) and co-workers,[72] and it reveals a very different history to that of the Sun and either of α Cen A or B (recall Figs. 3.10 and 3.13).

Strikingly, Proxima will not undergo a red giant phase, and rather than becoming larger and cooler as it ages, Proxima will become smaller and hotter. This result again comes about because of the full internal mixing that takes place within Proxima. As it

[72] See, G. Laughlin et al., "The end of the main sequence" (*The Astrophysical Journal*, **482**, 420, 1997). See also the highly recommended book by Fred Adams and Gregory Laughlin, *The Five Ages of the Universe: inside the physics of Eternity* (Simon and Schuster, New York, 1999).

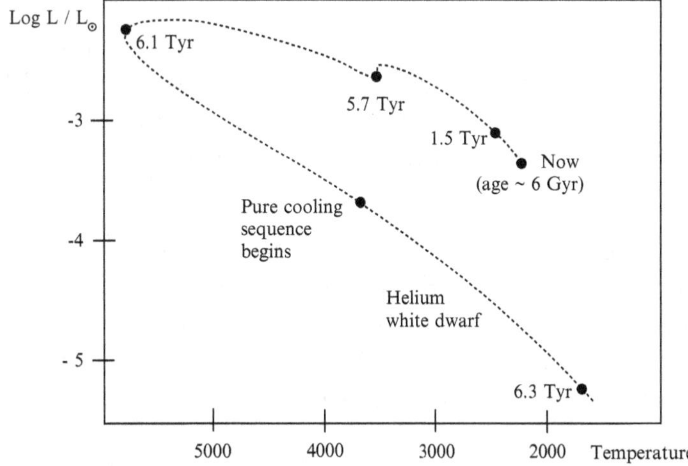

Fig. 3.15 The evolutionary track of Proxima Centauri. In contrast to the Sun, Proxima will not become a red giant; rather it will become smaller, hotter and more luminous as it moves through its main sequence phase. Upon exhausting hydrogen in its interior Proxima will enter into its white dwarf cooling phase, becoming slightly smaller but very much cooler and less luminous as time ticks by. Numbers indicate ages in trillions of years (Diagram based upon data and diagrams provided in G. Laughlin et al., "The end of the main sequence" (*The Astrophysical Journal*, **482**, 420, 1997))

ages Proxima certainly becomes more luminous, and this will change the boundaries of its habitability zone, but in terms of stable environments to prolong life (given an appropriately placed planet) Proxima and similar such M spectral-type red dwarf stars take the crown. Intriguingly, as well, the prospects of life evolving upon an appropriately located planet orbiting Proxima improve with time, since its flare activity, currently a potential life-destroying factor, will decay. Indeed, the observed flare activity among red dwarf stars appears to drop to near zero once they are more than seven to eight billion years of age. The evolution of life on a planet around a star like Proxima may be slow to appear, therefore, but it could potentially survive for an unimaginably long time.

Given the slow pace at which Proxima ages a hypothetical, hyper long-lived being, evolving in lockstep, would witness an incredible change in the cosmic vista. When Proxima eventually runs out of hydrogen to convert into helium the universe will be some 440 times older than at present, and indeed, the universe will then be entirely unrecognizable with respect to the one we see currently. About 1.5 trillion years from now Proxima will have advanced through just a sixth of its main sequence lifetime, and yet in this same timeframe the entire universe will have changed beyond all current recognition. Let the numbers speak for themselves:

- 4 billion years from now (YFN) – The great convergence of the Milky Way Galaxy and the Andromeda Galaxy begins, and any sentient beings (including our distant descendents – hopefully) will be able to witness two Milky Ways stretching across their night sky. The galaxy as we presently know and see it will be no more.
- 8 billion YFN – The Sun has become a raging red giant, and the destruction of Mercury and Venus is complete. The Solar System, as we currently know and see it, is no more.
- 10 billion YFN – Overrun by the expanding, post-main sequence envelop of α Cen B, the first discovered Centaurian planet α Cen Bb is destroyed.
- 15–20 billion YFN – The probability that a very close encounter with another star, thereby ejecting the remnant Earth from the white dwarf Sun's gravitational grasp, approaches certainty (see Fig. A3 in this book). The entire Solar System as we know it has now been destroyed.
- 20 billion YFN – The Big Rip end of the universe. In this extreme cosmological model the dark energy acceleration effect increases continuously with time, and space is ultimately ripped apart right down to the subatomic level. If this scenario holds up, and there is no present data to say that it won't, then Proxima will have a youthful and decidedly premature death.
- 100 billion YFN – Universal expansion carries all but the gravitationally bound Local Group galaxies[73] across the cosmic horizon, and apart from the view of local stars and nearby galaxies the universe begins its slow slide towards becoming a domain of

[73] The Local Group of galaxies extends over a region of about two to three million parsecs and contains perhaps as many as 50 galaxies. The Milky Way and Andromeda galaxies are the two dominant members of the group, and each has an attendant collection of many smaller satellite galaxies. The Local Group of galaxies currently forms part of the extensive Virgo Supercluster.

utter darkness. From this time onwards it will no longer be possible to show, by any experimental and/or observational means, that the universe is expanding. Indeed, the uniform expansion of space, one of the great cornerstones of modern observational cosmology, as first described by Edwin Hubble in the late 1920s, will no longer be evident or verifiable.

- 500 billion YFN – The Local Group of galaxies melds into one mega-galaxy, and the observable universe effectively becomes the single distribution of stars as modeled by Harlow Shapley in the early decades of the twentieth century.[74]
- 680 billion YFN – The cosmic background radiation is now undetectable, and the key signature betraying the hot Big Bang origin of the universe will no longer be knowable.[75] Indeed, it is

[74] Shapley's one large galaxy concept was essentially the same as that developed at the turn of the twentieth century by Jacobus Kapteyn. It was Kapteyn who began to use statistical parallax and proper motion data along with star magnitude counts to gauge the layout and shape of the stellar realm. Kapteyn's scheme was perhaps the first crowd-sourced science project, and he called upon astronomers from around the world to gather data in well- defined selected areas. Kapteyn also employed the labor of prisoners in a local jail to help with the reduction and analysis of photographic plates. The resultant profile Kapteyn deduced (circa 1910) was that of a single stellar conglomerate, about 40,000 light years across, having the shape of a flattened ellipsoid, with the Sun located at the center. The so-called Great Debate, which never actually took place, between Heber Curtis and Harlow Shapley in 1920, essentially determines the time at which astronomers began to agree that the universe is not composed of one large distribution of stars (as argued for by Shapley) but is, in fact, a vast arena of many independent galaxies or 'island universes' (as championed by Curtis). Once the Local Group of galaxies coalesce the ultimate profile, like the universes envisioned by Kapteyn and Shapley, will become that of a single superlarge elliptical galaxy, the whole embedded within a colossal, but otherwise star- and galaxy-devoid, cosmos.

[75] At this time the peak frequency relating to the cosmic background radiation will drop below the so-called plasma (also Langmuir) frequency of the interstellar medium. This condition specifically depends upon the number density of electrons in the interstellar medium. Once the frequency of the background radiation signal drops below the plasma limit it will no longer be able to propagate through the interstellar medium, and it will therefore be undetectable by any (mega-Local Group conglomerate) galactic observer. The steady erosion of cosmic history is nicely described by Lawrence Krauss and Robert Scherrer in their 2007 prize-winning essay, "The Return of a Static Universe and the end of Cosmology" (arXiv:0704.0221v3). Indeed, this essay was submitted to the annual competition organized by the Gravity Research Foundation in Wellesley Hills, Massachusetts. Founded by American economist Roger Babson in 1949, the institution was initially established with the aim of encouraging research into methods by which objects might be shielded from gravitational attraction. Had they actually been real, H. G. Wells's gravity-repelling paint cavorite (Note 22 above) and Isaac Asimov's gravitic drive (Note 23 above) would presumably have been of great interest to Babson.

now the case that every pillar upon which the modern theory of Big Bang cosmology is based will be either unknowable or unmeasurable.

- 1.5 trillion YFN – Star formation in the observable universe ends, there no longer being sufficient interstellar gas to form even the most diminutive of red dwarf stars. At this stage, Proxima is just 1.5 times more luminous than at present,[76] and it has barely evolved away from its formative zero-age main-sequence location in the HR diagram (recall Fig. 3.15).
- 6.1 trillion YFN – Proxima reaches its maximum luminosity, making it some 15 times more luminous than now (but still 160 times less luminous than the present-day Sun). Its surface temperature has now increased to 5,800 K, making it about 3,600° hotter than at the present time.
- 6.3 trillion YFN – Proxima finally exhausts the hydrogen within its interior and enters its helium white dwarf cooling phase. At this stage its luminosity and radius have dropped by factors of 8,000 and 10, respectively, and it is some 600° cooler than at the present time. The observable universe is now composed of the last formed faint red dwarf stars, white dwarfs, aged brown dwarfs and black holes. Hereafter, it begins to fade into near total darkness.

Having entered its white dwarf phase the nominal star-life of Proxima has ended. It will have long been stripped, by close star encounters, of any planets that might presently orbit around it, and its future is to slowly, mind-bendingly slowly, cool off to a temperature close to absolute zero. Over ensuing eons it will undergo close encounters with other degenerate white and black dwarf objects, but ultimately the end of Proxima, as indeed, the ultimate end phases of the remnant Sun, α Cen A, α Cen B, and all other stars, will be to suffer total disruption by black hole

[76] See, G. Laughlin et al., "The end of the main sequence" (*The Astrophysical Journal*, **482**, 420, 1997).

accretion – and at this time it effectively vanishes from our universe and beyond all knowledge.

The very far distant future is perhaps bleak from our human perspective, but it will be a very long time in coming, and between now and then it is at least reassuring to know that there are many deep adventures and many great journeys of discovery ahead of us.

"There are no secrets that time does not reveal."
– From the play *Britannicus* by Jean-Baptiste Racine (1639–1699)

Appendix I

1.1 The Magnitude Scale and Star Classification

The stellar magnitude scale is an anachronism – but it is, when all is said and done, a very important one. To the initiate student of astronomy it is an oddity in that a fundamental physical measurement is arbitrarily manipulated to give some other number. Indeed, behind the magnitude scale is a wealth of history and the acceptance of the fact that arbitrary schemes are useful, even if the reasons for the scheme are not immediately obvious.

Reducing observational astronomy to its bare bones, there are just three kinds of observations an astronomer can make. These are measurements of position, angular size and flux. All three of these measurements might vary with time, but the three fundamentals hold true – an astronomer can essentially measure where an object is in the sky, how big it looks through a telescope, and then how much energy per meter squared per second (the energy flux) is brought to the telescope's detector, be it through an eye, a CCD camera or a spectroscope. All that we know about the universe must be deduced from these three kinds of observation and their time variability.

Looking to the future these three basic measurements will be extended to include the detection of gravitational waves, and particle detectors such as cosmic ray detectors, neutrino labs and dark matter experiments – the latter only in recent times reaching the technological level at which directionality and source identification are possible.

For the magnitude scale it is the energy flux f measurement that is important. The greater the energy flux, in the form of electromagnetic magnetic radiation, the brighter an object will appear to the eye. With this being said, however, the relationship between flux and perceived brightness is a non-trivial one. Doubling the

M. Beech, *Alpha Centauri: Unveiling the Secrets of Our Nearest Stellar Neighbor*, Astronomers' Universe, DOI 10.1007/978-3-319-09372-7,

flux does not mean that an object will appear twice as bright. Although the flux measurement is (in principle) a direct account of the energy received at Earth's orbit from a star, it does not tell us anything fundamental about the star's actual energy output (its luminosity L). To determine this latter quantity, knowledge of a star's distance d is additionally required.

The flux, luminosity and distance are related through an inverse square law, with $f = L/4\pi\, d^2$. For a star close enough to the Solar System that a parallax measurement can be made, the distance d can be determined directly, and this enables the luminosity, literally the amount of electromagnetic energy radiated by the star into space (assumed isotropically at this stage) per second (the *SI* units are watts or joules per second), to be derived. The luminosity is a fundamental property of a star and it tells astronomers about the energy generation processes that must operate within stellar interiors.

Not all stars have distances that are well known, however, and accordingly the flux measurement is all that astronomers have to work with – which is not to say that there isn't a vast amount of information encoded within the flux measurement. Astronomers working in the more contemporary areas of observational astronomy, using radio, microwave, X-ray and UV telescopes, express their flux measurements directly in terms of the measured value at a specific wavelength – what they measure is what they print. Astronomers working at optical and infrared wavelengths, however, have traditionally used the magnitude scale to express their measurements of flux. Where and why, it is now reasonable to ask, did all this tinkering with the flux measurements begin?

The modern story begins in the mid-nineteenth century with the work of British astronomer Norman Pogson.[1] In grand Victorian fashion, Pogson was interested in standardization, and specifically standardization of brightness estimates. What he wanted to establish was a system, arbitrary as it might be, that everyone agreed to use and which would establish an instrument-based standardized system for brightness measurements. He built upon

[1] See the very readable and informative article by John Hernshaw (University of Canterbury, New Zealand), "Origins of the stellar magnitude scale" (*Sky & Telescope*, November 1992).

two earlier, historical observations. First, the classic catalog of stars compiled circa 135 B.C. by Greek astronomer Hipparchus described star brightness in terms of six magnitude categories. Again, this division of six categories was entirely arbitrary, and Hippachus could have chosen five or seven, or ten magnitude intervals. History tells us, however, Hipparchus choose five magnitude categories, with magnitude 1 stars being the brightest and magnitude 6 stars being the faintest visible to the human eye. Such a system is perfectly fine, and it simply provides an array of pigeonholes into which stars of varying brightness can be sorted – so much for ancient history. Moving forward two millennia, however, the great William Herschel noted from his pioneering measurements that the flux recorded for a first magnitude star was approximately 100 times larger than the flux that was recorded for a sixth magnitude star. It was this observation that appealed to Pogson, and he realized that the ancient and modern magnitude and flux measurements could be brought into agreement if the magnitude m is defined as being related to the logarithm of the flux:

$$m = -2.5\log(f) + C \qquad \text{(A.1)}$$

where C is a constant to be determined. The 2.5 constant term that multiples the logarithm appears in Eq. A.1 in order that the Hipparchus magnitude scheme and the observational discovery of Herschel are in agreement. Indeed, if we have two stars of magnitude m_1 and m_2 with corresponding measured flux values f_1 and f_2, then applying Eq. A.1 twice and subtracting we find

$$m_1 - m_2 = 2.5\log(f_2 / f_1) \qquad \text{(A.2)}$$

Now, if m_1 is a sixth magnitude star and m_2 is a first magnitude star, we see from Eq. A.2 that $m_1 - m_2 = 5 = 2.5 \log (f_2/f_1)$, from which we find that $f_2/f_1 = 10^{5/2.5} = 100$, and this is exactly the observational result found by Herschel. The flux received from a first magnitude star is 100 times greater than that received from a sixth magnitude star. Alternatively, by way of another example, if the fluxes measured for two stars differ by a factor of 25, then they also differ by $2.5 \log (25) \approx 3.5$ magnitudes.

To the Victorian mindset this harmonization between ancient and modern methods was beautiful, and astronomers, after an initial few decades of delay, eventually adopted Pogson's magnitude scheme. Having established a method for finding the magnitude difference between any two stars, the system can now be extended to objects fainter than magnitude +6, the erstwhile naked-eye limit. In general the larger and more positive the magnitude the fainter a star is, and modern instruments, such as the Hubble Space Telescope, can detect stars to an apparent magnitude of about +25. The Sun, the brightest observable object in the entire sky, has an apparent magnitude of –26.72.

Pogson's method is more than just semantics and organization, however, and its real power lies in the fact that by defining a list of standard stars, it enables the establishment of an instrument-based system for reporting observational results. Indeed, it is essentially through the adoption of standard stars, of specific and agreed upon magnitudes, that the constant C is determined in Eq. A.1.

In this manner any new telescope and flux measuring system can be calibrated, and all observers, no matter when and/or where they are observing from and with whatever equipment design, will report consistent magnitude values. For Pogson specifically, as a pioneer of variable star investigations, and for astronomers in general, this consistency is of primary importance, since it now means that any variations reported in the magnitude value for a specific object are real, and not the result of some quirk of the astronomer's eye or the instrumentation. The scheme might be arbitrary, but it is consistent and well defined, and it enables astronomers to determine real time variations in the energy flux from astronomical objects.

The apparent magnitude of a specific star will vary according to its distance from the Solar System. This follows from the $f = L/4\pi d^2$ relationship. By introducing the absolute magnitude M as the apparent magnitude a star would have if viewed from a distance of 10 pc (again an arbitrary but reasonably chosen distance value), we can derive from Eq. A.2 the useful result that

$$m - M = 5\log(d/10) \tag{A.3}$$

this equation enables the determination of the apparent magnitude m of a star for a given distance d. Or vice versa if the absolute magnitude is appropriately calibrated, so a measure of the apparent magnitude can be used to determine the distance to a star. The latter, so-called standard candle, method of distance determination works provided the absolute magnitude has been calibrated against the observable spectral characteristics of the star. For Sun-like stars, $M = +4.83$, while for stars like Proxima, $M = +15.48$.

In the modern era the flux measurement and magnitude scheme has been developed to a very high level of sophistication, and fluxes are measured through very specific and carefully manufactured sets of transmission filters.[2] The most commonly used filter system is that introduced in the 1950s by Harold Johnson and William Morgan. In this UBV filter scheme astronomers measure the flux from a given star through each of the three filters in turn and then construct what are called color indices from the deduced magnitudes. These color indices can then be used to reveal additional fundamental data about a star; the (B-V) color index, for example, is directly related to the temperature of star through the relationship $T \approx 7{,}000/[(B\text{-}V) + 0.7]$.

The power of the standard candle method, and the utility of Eq. A.3, relies entirely on calibration and being able to recognize similar stars on a distance-independent basis. This latter requirement is achieved through the study of stellar spectra – literally, the spreading out of starlight into its rainbow colors.

For all of the poetical beauty associated with the rainbow, however ("My heart leaps up when I behold a rainbow in the sky" are Wordsworth's feelings), a stellar spectrum, as far as the astronomer is concerned, is simply a measure of the energy flux received at the telescope's detector as a function of the wavelength of light. The energy flux is typically measured in the optical part of the spectrum – from blue light with $\lambda \sim 450$ nm to red light with $\lambda \sim 750$ nm. A stellar spectrum is a composite of a continuous spectrum, which provides some energy flux at all wavelengths, and absorption lines. The continuum spectrum is close to the theoretical ideal of a blackbody radiator, first successfully described by

[2] Although now a little dated, the classic text, still very much worth reading, is that by Jean Dufay: *Introduction to Astrophysics – the Stars* (Dover Publications Inc., 1964).

German physicist Max Planck towards the close of the nineteenth century. Accordingly, the maximum flux will fall at a wavelength λ_{max} described by Wiens law: $\lambda_{max}\,T = 2.8977721 \times 10^{-3}$ mK, where T is the surface temperature of the star. It is the characteristic surface temperature that determines the color of a star. Cool stars will have λ_{max} towards the long, red wavelength end of the visual spectrum; hot stars will have λ_{max} towards the blue.

Superimposed on the continuum spectrum of a star are absorption lines at very specific wavelengths. These lines are the result of a reduced continuum flux reaching the telescope's detector – the consequence of atoms and in some cases molecules in a star's photosphere robbing continuum photons to enable electron transitions. Not only do the wavelength locations of the absorption lines betray the kinds of atoms (and possibly molecules) that exist in the outer layers of a star, the strength of the lines– that is, how wide and deep they are – can used to establish a spectral classification scheme.

The standard spectral classification scheme recognizes seven main types, signified O, B, A F, G, K and M. Each type is further divided into subgroups indicated by a numerical digit between 0 and 9.[3] The spectral type is a distance-independent proxy measure of star temperature, with the O stars being the hottest, having temperatures in excess of 30,000 K. The coolest, M spectral-type stars, in contrast, have temperatures as low as 2,500 K. The spectral designation that is assigned to a star is distance-independent since the scheme uses absorption line strength ratios to specify the spectral type and subgroup number.

In addition to the spectral type a second classification can be determined according to line strength characteristics. This provides the so-called luminosity class, and its value is indicated according to a Roman numeral that runs from I to V. The I luminosity class is further divided into Ia and Ib, and this applies to the most luminous supergiant stars.

The Sun is a luminosity class V (also called main sequence) star. The luminosity class is partly a proxy measure of a star's surface gravity. Supergiant, luminosity class I, stars, with large and extended outer atmospheres, produce narrow and well-defined

[3] A detailed discussion of the various spectral classification schemes presently in use is given by Perry Berlind at www.cfa.harvard.edu/~pberlind/atlas/htmls/atframes.html.

TABLE A.1 Magnitudes and distance values for the principal stars in Centaurus. Details are also shown for Proxima Centauri. Apparent magnitude and spectral-type data for the principal stars are from the *Bright Star Catalog* and the distances are from SIMBAD

Star	Apparent Magnitude	Absolute Magnitude	Distance (parsecs)	Spectral Type	Luminosity Class
α	-0.27	4.082	1.348	G2+K1	V+V
β	0.60	-4.799	120.192	B1+ B1	III+III
γ	2.17	-0.481	33.904	A1+A1	IV+IV
δ	2.57	-0. 260	127.227	B2	IV
ε	2.34	-3.247	131.062	B1	III
ζ	2.55	-2.793	117.096	B2	IV
θ	2.06	0.780	18.034	K0	III
ι	2.73	1.451	18.020	A2	V
Proxima	11.05	15.54	1.300	M5	V

absorption lines, while, at the same temperature (that is, spectral type), more compact luminosity class V dwarf stars have relatively broad lines because of pressure broadening effects.

Although not really stars, since they are not massive enough to enable the initiation of sustained hydrogen fusion reactions, three additional letters have recently been added to the seven standard classification letters to cover the brown dwarfs. Again, the letters cover a range of temperatures, with the L spectral type accounting for the hottest brown dwarfs with temperatures varying from 1,300 to 2,500 K (this technically allows for some overlap with the coolest M spectral-type stars).

Spectral types T and Y cover the lowest temperature range of the brown dwarfs. The T brown dwarfs have temperatures in the range 1,300–700 K, and the Y spectral type covers the lowest temperature range. The brown dwarf, as of this writing, with the lowest recorded temperature was discovered by the Wide field Infrared Survey Explorer (WISE) spacecraft in 2011, and WISE 1828+2650 (a Y2 brown dwarf) has a temperature of about 400 K – just above the temperature of boiling water at sea level on Earth. The Y spectral class bridge the temperature gap between the coolest brown dwarfs and Jovian planets.

The magnitude, distance, spectral type and luminosity classes for the principal stars in Centarus are given in Table A.1. A range of star types is apparent, with ε Centauri being the hottest star, θ Centauri being the largest, β Centauri (Agena) being the most luminous, with Proxima holding the smallest and lowest temperature star spot.

Of the stars listed in Table A.1 three are in binary systems. Here we just describe β and γ Centauri – since α Cen AB is described extensively within the text. The 'star' Agena (β Centauri) was only identified as being a binary system in 1999, with its full orbit only becoming known in 2005, following the analysis by John Davis (University of Sydney) and co-workers. The two stars have nearly identical masses of 9.09 times that of the Sun, and they orbit each other once every 357.0 days. At their closest approach they are just 0.5 AU apart. They are 5.5 au apart when at their greatest separation. The binary status of Muhlifain (γ Cen AB) was first discussed by John Herschel in 1835, although John Ellard Gore made the first orbital determinations in 1892 – the star was also studied by Innis and Voute in the early 1900s. The two near identical three solar mass stars orbit each other once every 84.5 years.

An extremely useful way of showing the differences between stars is to plot them in a Hertzsprung-Russell (HR) diagram. Indeed, this diagram provides information about a star's temperature, radius and luminosity – the diagram's axis being a combination of these factors usually expressed in terms of spectral type, or (B-V) color index, and the star's luminosity is typically expressed as its absolute (or apparent) magnitude. The regions in which the different luminosity classes plot in the diagram further indicate the size of a star. Statistical studies indicate that of order 92 % of all stars are of luminosity class V, and such stars fall on the main sequence in the HR diagram. It is these main sequence stars that are converting hydrogen into helium within their central cores. Low mass, M-spectral type, luminosity class V stars (also called red dwarfs) plot in the lower right hand corner of the HR diagram and are located at the bottom of the main sequence. Massive, O spectral-type, luminosity class Ib stars plot in the upper left hand corner of the HR diagram at the top the main sequence. Giant and supergiant luminosity class III, II and Ia stars plot above the main sequence diagonal and towards the upper right hand corner in the HR diagram. The giant stars are generating energy by converting helium into carbon within their central cores. Giant and supergiant

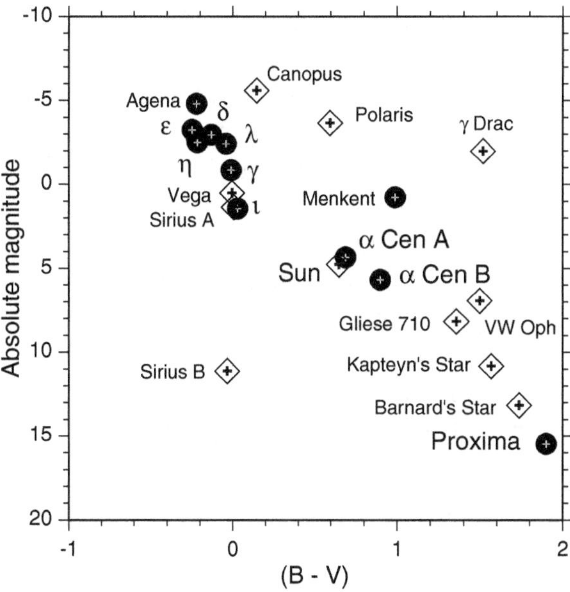

FIG. A.1 An HR diagram of the principal stars in Centaurus and a collection of additional (comparative) stars. The (B-V) X-axis, indicates the temperature of a star. The absolute magnitude, Y-axis, is a measure of the star's luminosity. The radii of the star increases as it moves from the lower left hand corner of the diagram to the upper right hand corner – the radius being related to the temperature and luminosity through the Stefan-Boltzmann formula

stars are not common, and only about 1 % of all the observed stars, at the present time, fall into these specific groups. The location of a star in the HR diagram is essentially described by the Stefan-Boltzmann relationship for blackbody radiators. This formula links the luminosity L to the radius R and temperature T as

$$L = 4\pi R^2 \sigma T^4 \tag{A.4}$$

where $\sigma = 5.6697 \times 10^{-8}$ Wm^{-2} K^{-4} is the Stefan-Boltzmann constant.

Figure A.1 shows the HR diagram for the stars in Centaurus, along with Proxima and a few well-known and famous stars (as well as some of the stars that are specifically mentioned within the main text). The North Star (Polaris = α Ursae minoris[4]), for example, is shown in the diagram, and it falls in the giant region

[4] Technically Polaris is part of a triple-star system and has the designation α UMi Aa.

corresponding to luminosity class Ib. Menkent (θ Centauri) also plots in the giant region, according to its luminosity class III designation. Agena (β Centauri) is the most luminous, most massive, highest temperature main sequence star in the Centaurus group. The star Sirius A (α Canis Majoris), while less luminous than Agena, is nonetheless the brightest star in our sky because it is 45 times closer to us. The white dwarf companion to Sirius A, the star Sirius B, plots below the main sequence, indicating a low luminosity and small radius for its temperature (as required by the Stefan-Boltzmann equation). The Sun-like properties of α Cen A and α Cen B are revealed by plotting close to the Sun's location in the HR diagram – all three stars being main sequence luminosity class V stars.

Appendix 2

1.1 Stellar Motion and Closest Approach

To determine the path of a star through space a measure of its distance d, proper motion μ and radial velocity V_R at some specific epoch are needed.[5] These data provide information about the speed and location at some fixed point at some specific time, and it is relative to this reference position that all others, past and future, are referred. The distance to a star can be measured directly from its parallax (see Sect. 1.10); likewise the proper motion can be directly measured, and this quantity indicates the observed angular rate of motion across the sky, in units of arc seconds per year.

The radial, or line-of-sight, velocity of a star is further measured directly by exploiting the Doppler effect. This latter phenomenon, first described in detail by Christian Doppler in 1842, is most familiar to us through its effect upon sound waves emitted by moving objects, such as emergency response vehicles. The Doppler effect describes the pitch variation, increase followed by a decrease, that we hear when a sound source first approaches and then recedes from our fixed location. What is taking place is not a change in the sound wave emitted but rather a modification of the sound wave that is heard. When the object is moving towards us the wavelength that we hear is perceived to be shorter than the wavelength of the sound emitted, and in contrast the wavelength

---

[5] The methodology behind calculating the time-varying distance to nearby stars is given by Jocelyn Tomkin in *Sky & Telescope* magazine for April 1998. More detailed methodologies and calculations are presented by R. A. J. Mathews in the *Quarterly Journal of the Royal Astronomical Society*, **35**, 1 (1994), and by V. V. Bobylev in the journal *Astronomy Letters*, **36**, 220 (2010). The latter author considers the full three-dimensional galactic model, equations of motion for the stars and also looks closely at the possible encounter conditions for GL 710, finding that there is an 86 % chance that the star will cut inside of the Oort Cloud boundary in the time interval 1.45 ± 0.06 million years hence. There is also a non-zero, but very small probability (10^{-2} percent, or 1 in 100 chance), that GL 710 will pass as close as 1,000 au from the Sun, where in principle it could not only influence the Solar System's Oort Cloud but its Kuiper Belt as well.

M. Beech, *Alpha Centauri: Unveiling the Secrets of Our Nearest Stellar Neighbor*, Astronomers' Universe, DOI 10.1007/978-3-319-09372-7,

that we perceive when an object is moving away from us is longer than that emitted. The variation in the perceived wavelength is directly related to the speed with which the sound source is moving.

Although sound waves provide an everyday example of the Doppler effect, astronomers can also exploit the effect by looking for wavelength variations in the light received from a star. In the case of a star, if a specific feature (such as an absorption line) that should, if everything was at rest, be observed at some known wavelength λ, is observed at some other wavelength λ_{obs}, then the shift $\Delta\lambda = \lambda_{obs} - \lambda$ is related to the speed V_R of the star in the observer's line-of-sight (hence radial velocity) via the expression

$$\frac{\Delta\lambda}{\lambda} = \frac{V_R}{c} \tag{A.5}$$

where c is the speed of light.

Since the observed shift can be either positive or negative, the convention is that a positive velocity indicates motion away, whereas a negative velocity indicates motion towards the observer. Astronomers tend to describe these motions in terms of the starlight feature being either red-shifted or blue-shifted. This simply expresses the increase or decrease in the observed wavelength and the direction in the color spectrum (between red and blue) towards which it is shifted.

Figure A.1 shows the essential geometry of the star motion problem. Key to making progress is the determination of V_S, the true spatial velocity of the star through space. This can be done through the construction of what is called the tangential velocity V_T, which is expressed in terms of measured proper motion and parallax. Indeed, by construction we have: $V_T = 4.74\,(\mu/\pi)$ where π is the measured angle of parallax in arc seconds and the distance to the star, in parsecs, is $d = 1/\pi$. The odd-looking constant of 4.74 reflects the units being used for the distance and proper motion and gives the tangential velocity in kilometers per second. With the tangential velocity and radial velocity determined the space velocity of the star follows from the Pythagorean rule, with $V_S = \sqrt{V_R^2 + V_T^2}$. In addition we can also determine the direction θ with which the star is moving relative to the line of sight through

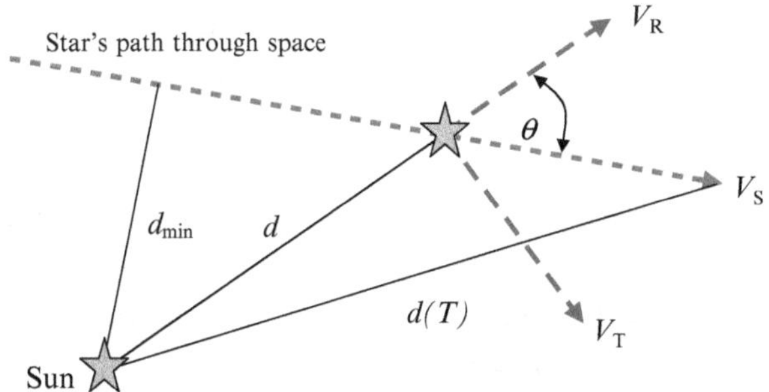

FIG. A.2 The motion of the star is described in terms of its space velocity V_S and the angle θ that it makes to the line of sight along which the radial velocity V_R is measured via the Doppler effect

the relationship tan $\theta = V_T/V_R$. The location of the star $d(T)$ at any time T into the past or into the future (relative to the time at which the parallax and proper motion were measured) can now be determined through the cosine rule

$$d(T) = d^2 + (TV_S)^2 - 2dTV_S \cos(180 - \theta) \qquad \text{(A.6)}$$

The distance of closest approach d_{min} is determined, again with reference to Fig. A.2, through the relationship, $d_{min} = d \sin \theta$ and the time T_{min} since, or, depending upon the direction of motion, the closest approach point will be $T_{min} = d \cos \theta / V_S$. Combining Eq. A.5 with Eq. A.3 further enables the determination of the apparent magnitude of the star as it moves towards or away from the Solar System.

The actual motion of a star through space, around the galactic center, is more complicated than that given in the discussion here and as illustrated in Fig. A.2.[6] The spatial motion is really three dimensional, and there is no long-term guarantee that the space

[6] Technically six parameters are required to determine a star's future trajectory through space: these are its galactic (x, y, z) position coordinates and its galactic (Vx, Vy, Vz) velocity components. Once these parameters are specified at one specific epoch then, adopting some specific model for the gravitational potential of the Milky Way galaxy (technically including the central bulge, the disk, as well as the halo), the equations of motion can be numerically integrated forwards (or backwards) in time. An example of such calculations is presented by Juan Jiménez-Torres et al., in the research article "Effect of different stellar galactic environments on stellar discs – I: The solar neighborhood and the birth cloud of the Sun", published in *The Monthly Notices of the Royal Astronomical Society* (**418**, 1272, 2011).

velocity will remain constant, since stars are continually affected by the gravitational forces relating to other nearby stars.

The typical time T_{enc} between encounters, at a miss distance r, between the Sun and another star can be estimated from the geometrical cross-section area πr^2, and the typical speed V_S with which stars move within the solar neighborhood. In this manner, the cylindrical volume U swept out by the Sun in time T_{enc} will be $U = \pi r^2 V_S T_{enc}$. The number of stars N^* that will be encountered while sweeping out this volume of space will then be $N^* = \rho^* U$, where $\rho^* = 0.09$ is the number density of stars per unit parsec cubed of space. Taking now the encounter time to be the time required for one star to be encountered (that is $N^* = 1$) within the cylindrical volume we have

$$T_{enc} = \frac{1}{\rho^* \pi r^2 V_S} \tag{A.7}$$

putting in typical values and expressing T_{enc} in years, V_S in kilometers per second and r in parsecs, we find that $T_{enc} \sim 150{,}000$ years when $r(\text{pc}) = 1$ and $V_S = 25$ km/s. This is the typical expected encounter time interval between stars passing as close as α Centauri to the Sun. Equation A.7 reveals another interesting point, in that the encounter time interval increases dramatically as the encounter miss distance decreases. This indicates that direct collisions between individual stars must be very rare within the galaxy, since a direct collision requires $r(\text{pc}) \sim 0$. Close encounters such as that expected between the Sun and Gliese 710 in about 1.6 million years (see Sect. 1.16), with a miss-distance of about 0.3 parsecs, should occur on timescales of order once every three million years or so. Alternatively, we can calculate that the closest approach of another star to the Sun since it formed 4.56 billion years ago, is likely to have been no smaller than about 2,000 au – a distance well within the Oort cometary cloud, but not one close enough to disrupt (as we would hope) planetary orbits. Figure A.3 shows the variation of T_{enc} against miss-distance r in the range 10^{-4} to 1 pc.

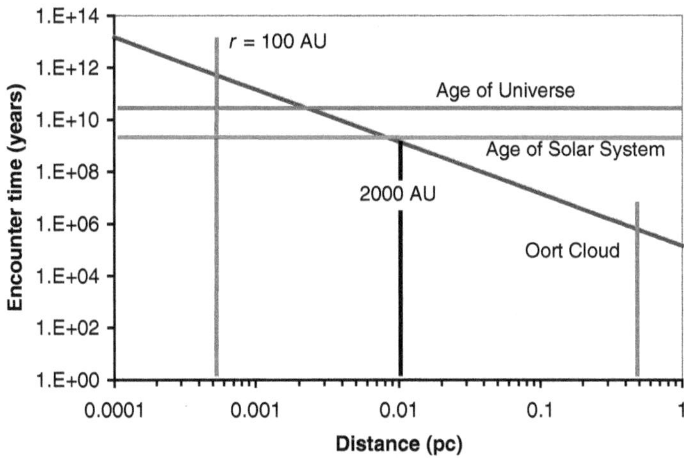

FIG. A.3 T_{enc} versus miss-distance r as given by Eq. A.7 when $V_S = 25$ km/s. The two horizontal lines correspond to interval times given by the age of the Universe (13.7 billion years) and the age of the Solar System (4.56 billion years). The time interval between encounters closer than $r = 100$ au is currently greater than the age of the universe, and since the Solar System formed no encounter closer than $r \sim 2{,}000$ au is likely to have happened. The time interval between successive passages of a star through the outer boundary of the Oort Cloud ($r \sim 10{,}000$ au) is order 665,000 years

Appendix 3

1.1 The Orbit and Location of α Cen B

A total of seven parameters are required in order to describe the sky position of α Cen B relative to α Cen A. These parameters are summarized in Table A.2. The basic shape of the orbit is determined by the size of the semi-major axis a, and the eccentricity e. The orientation of the orbit to our line of sight is additionally described by the inclination i, the longitude of the ascending node Ω, and the argument of periastron Ω. The final parameter T_0 provides the time during which the last periastron passage of α Cen B occurred.

Without going into specific derivations,[7] the equations that describe the position of α Cen B at any time t relative to the time of periastron passage T_0, are:

$$M = \frac{2\pi}{P}(t - T_0) \tag{A.8}$$

$$M = E - e \sin E \tag{A.9}$$

$$r = a(1 - e \cos E) \tag{A.10}$$

where M is the so-called mean anomaly, E is the eccentric anomaly, P is the orbital period, a is the semi-major axis, e is the eccentricity and r is the radial distance of α Cen B from α Cen A. Equation A.9 corresponds to what is known as Kepler's equation, and it has no simple analytic solution (except when $e=0$, which corresponds to the case of a perfectly circular orbit). Accordingly an iterative solution to Eq. A.9 must be found, but this can be done

[7] A detailed review of Keplerian orbits and dynamics, and specifically the solution to Kepler's problem, is provided by C. D. Murray and A. C. M. Correia – see arXiv:1009.1738v1.

M. Beech, *Alpha Centauri: Unveiling the Secrets of Our Nearest Stellar Neighbor*, Astronomers' Universe, DOI 10.1007/978-3-319-09372-7,

TABLE A.2 Orbital parameters (and their associated uncertainties) for α Cen B. The data is taken from Pourbaix et al.[a]

Element	Value
a (arc seconds)	17.57 ± 0.02
e	0.5179 ± 0.0008
i (degrees)	79.20 ± 0.04
Ω (degrees)	231.65 ± 0.08
Ω (degrees)	204.85 ± 0.08
T_0 (year)	$1{,}875.66 \pm 0.01$
P (years)	79.91 ± 0.01

[a]D. Pourbaix *et al.*, "Constraining the difference in convective blueshift between the components of α Cen with precise radial velocities" (*Astronomy & Astrophysics*, **386**, 280, 2002)

numerically using the Newton–Raphson technique in which an iterative solution for E is found via the scheme

$$E_{j+1} = E_j - \frac{E_j - e\sin E_j - M}{1 - e\cos E_j}, \quad j = 0,1,2,3,.. \tag{A.11}$$

The iteration scheme in Eq. A.11 starts by assuming some value for E_0, any reasonable value will do, so say start with $E_0 = 0.25$. With this starting value for $j = 0$, use Eq. A.11 to determine E_1, and with E_1 use Eq. A.11 again to determine E_2 and so. Keeping track of the various E_j terms will reveal that they converge to a constant value as j becomes larger and larger. Once the difference between successive iterations for E_j become smaller than say 10^{-6}, a solution is said to have converged. This is the value of E that is then used to determine the radial distance r via Eq. A.10 of α Cen B from α Cen A.

Rather than using the eccentric anomaly to describe the trajectory of α Cen B, it is generally easier to use the true anomaly f, which can be determined from Eq. A.9 through the relationship

$$\cos f = \frac{\cos E - e}{1 - e\cos E} \tag{A.12}$$

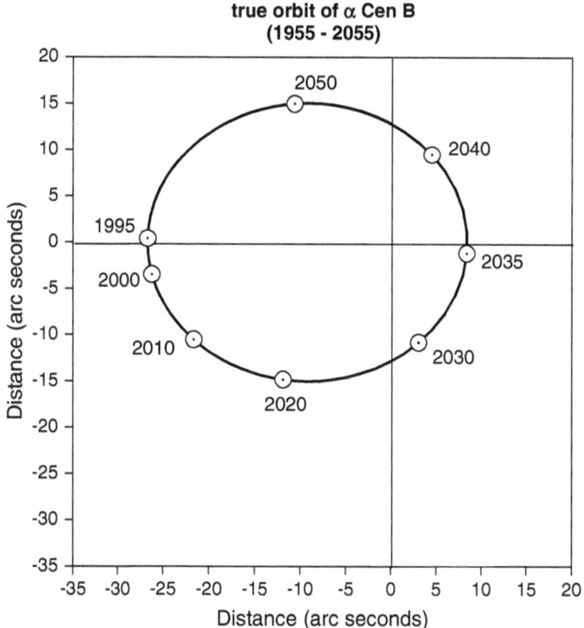

FIG. A.4 The true orbit of α Cen B about α Cen A. The position of α Cen B is shown for January 1 for various years. The scale is in arc seconds, and α Cen A is located at the (0, 0) point

Equations A.10 and A.12 determine the specific location of α Cen B in its elliptical orbit (the so-called true orbit) at time t. Figure A.4, shows the location of α Cen B at various times in the time interval between 1955 and 2055.

Thus far we have shown how to find the position of α Cen B relative to α Cen A with respect to its true orbit. From Earth, however, we do not see the true orbit; rather we see the apparent orbit – the latter being different since it includes the geometric orientation of the orbit in our line of sight. The apparent orbit is derived from the true orbit (Fig. A.4) by introducing rotation transformations relating to the orbital inclination (i), the argument of periastron (Ω) and the longitude of the ascending node (Ω). These angles are given in Table A.1. First, the three-dimensional (that is, true spatial) orbit can be constructed in the X, Y and Z coordinate frame, where the X-axis lies along the orbital major-axis, the Y-axis is at right-angles to the X-axis and the Z-axis is perpendicular the XY plane. The equations that describe the (X, Y, Z) positions are[8]:

$$X = r\left(\cos\Omega\cos(\omega + f) - \sin\Omega\cos(\omega + f)\cos i\right) \tag{A.13}$$

$$Y = r\left(\sin\Omega\cos(\omega + f) + \cos\Omega\sin(\omega + f)\cos i\right) \tag{A.14}$$

$$Z = r\sin(\omega + f)\sin i \tag{A.15}$$

Although Eqs. A.13, A.14 and A.15 provide a description of the orbit in three dimensions, the stars are physically observed on the two-dimension plane of the sky, and accordingly a polar coordinate system is introduced. In this case the position, in our case, of α Cen B relative to α Cen A, is described according to a position angle θ, and a radius offset ρ (see the inset in Fig. A.5). The position angle is measured in degrees from the north position and the radial offset, or separation, is given in arc seconds. The equations for describing the (θ, ρ) polar coordinate positions are:

$$\tan(\theta - \Omega) = \tan(\omega + f)\cos(i) \tag{A.16}$$

$$\rho = r\cos(\omega + f) / \cos(\theta - \Omega) \tag{A.17}$$

Figure A.5 shows the time variation of the position angle and the separation of α Cen B in the time interval of 1995–2075. From Fig. A.5 we see that α Cen B will next be due north of α Cen A (θ = 0°) in the year 2023, when the two stars will be separated by about 7 arc sec. The two stars will be at their closest approach, just 2 arc sec apart, in 2038; their greatest angular separation, about 22 arc sec, will be achieved in 2060. Table A.3 indicates the position angle and offset separation for January 1 in each year from 2013 to 2023.

[8] A detailed review of Keplerian orbits and dynamics, and specifically the solution to Kepler's problem, is provided by C. D. Murray and A. C. M. Correia – see arXiv:1009.1738v1.

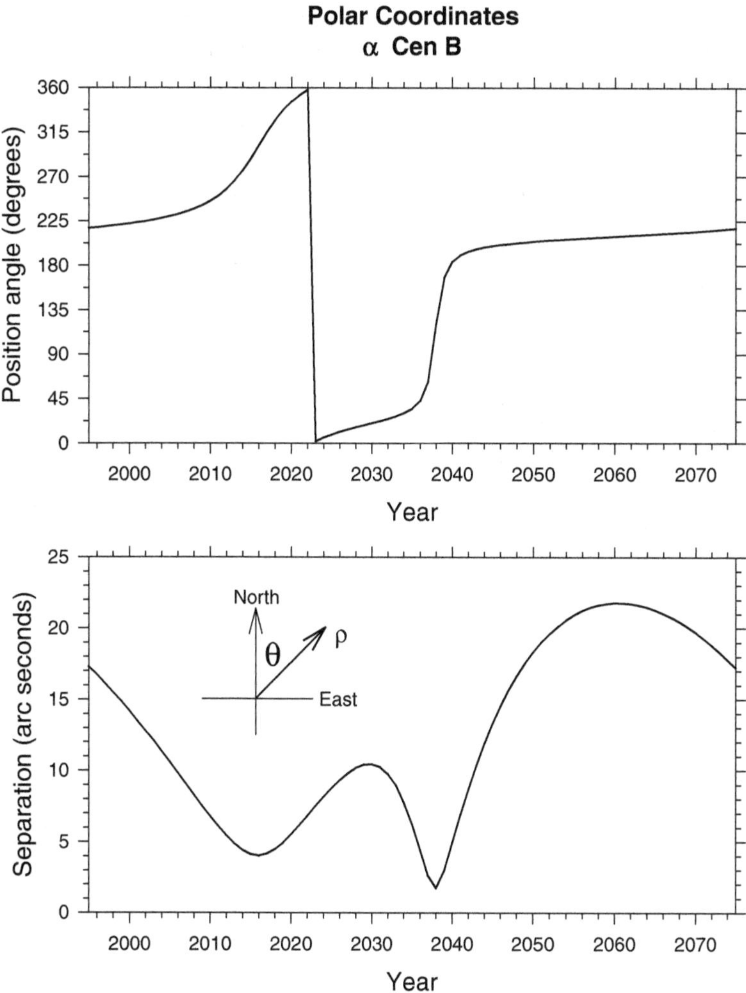

FIG. A.5 Time variation of the position angle (*top* panel) and separation (*bottom* panel) of α Cen B from α Cen A. The *inset* diagram on the bottom panel describes the polar coordinate system

Table A.3 Position angle (in degrees) and offset separation (in arc sec) of α Cen B from α Cen A on January 1 for the years indicated in Column 1

Year	Position Angle (°)	Separation (arc sec)
2013	265.8	4.9
2014	276.2	4.4
2015	288.5	4.1
2016	302.0	4.0
2017	315.4	4.2
2018	327.5	4.5
2019	337.6	4.9
2020	345.8	5.5
2021	352.4	6.1
2022	357.8	6.8
2023	2.1	7.5

Index

M. Beech, *Alpha Centauri: Unveiling the Secrets of Our Nearest Stellar Neighbor*, Astronomers' Universe, DOI 10.1007/978-3-319-09372-7,

CPSIA information can be obtained
at www.ICGtesting.com
Printed in the USA
LVHW061657060421
683579LV00016B/946

9 783319 093710